Series / Number 07-007

ANALYSIS OF NOMINAL DATA

SECOND EDITION

H. T. REYNOLDS
University of Delaware

SAGE PUBLICATIONS
The International Professional Publishers
Newbury Park   London   New Delhi

For information address:

 SAGE Publications, Inc.
2455 Teller Road
Newbury Park, California 91320
E-mail: order@sagepub.com

SAGE Publications Ltd.
6 Bonhill Street
London EC2A 4PU
United Kingdom

SAGE Publications India Pvt. Ltd.
M-32 Market
Greater Kailash I
New Delhi 110 048 India

Printed in the United States of America

International Standard Book Number 0-8039-0653-6

Library of Congress Catalog Card No. 77-72851

03   04   20   19   18   17   16   15   14   13

When citing a University Paper, please use the proper form. Remember to cite the correct Sage University Paper series title and include the paper number. One of the following formats can be adapted (depending on the style manual used):

(1) REYNOLDS, H.T. (1984) *Analysis of Nominal Data* (2nd ed.). Sage University Paper Series on Quantitative Research Methods, Vol. 7. Newbury Park, CA: Sage.

or

(2) Reynolds, H. T. (1984). *Analysis of nominal data* (2nd ed.) (Sage University Paper Series on Quantitative Research Methods, Vol. 7). Newbury Park, CA: Sage.

CONTENTS

Series Editor's Introduction to the Second Edition

Professor H. T. Reynolds has written a second edition of *Analysis of Nominal Data* at our urging, in order to update the presentation and to fit it more carefully into the entire package of analysis techniques offered in this series. In this edition he presents an expanded discussion of the chi square test of significance and its use in analyzing nominal, or categorical, data. At the same time, he introduces the reader to the many measures of association available for use in tables containing only categorical data. This discussion is particularly clear, and the logic underlying each measure is nicely spelled out for the beginning reader. Professor Reynolds starts at the beginning, explaining concepts in such a fashion that the novice can follow the presentation easily. That he is able to accomplish this without sacrificing rigor is a tribute to his abilities as a scholar and an author.

Professor Reynolds compares the myriad potential measures of association for tables of categorical data largely in terms of their definition of "perfect association": as strict, as implicit, or as weak perfect association. In strict association, each value of one variable is uniquely associated with a value of the other. In implicit perfect association, members of a column classification are "as homogeneous as possible" with respect to the row variable in the sense that there is only one nonzero row entry per column. Finally, weak perfect association refers only to the situation in which the categories of one variable are as homogeneous as possible with respect to the second, *given* the limitations imposed by differing marginal totals. Various measures of association differ in this respect, and also in the types of tables to which they are most aptly applied. Professor Reynolds makes useful recommendations on each of these matters.

In the first edition, Professor Reynolds provided extensive coverage of the multivariate analysis of nominal data. He devoted over 25 pages of an 80-page monograph to this topic. Since the first edition was published, however, this series has published a monograph on log linear models, by David Knoke and Peter Burke. As a result, Professor Reynolds has cut back on his presentation of multivariate methods, and altered the remaining discussion to serve primarily as an introduction to the topic for readers who intend to follow up the topic by reading the Burke and Knoke volume. He introduces the concept of a cross-product ratio, also called the odds ratio, and discusses its properties and how it can be used in the analysis of contingency tables. He demonstrated its use in both 2×2 and $I \times J$ tables.

Finally, I might point out that Professor Reynolds's discussion of Proportional Reduction in Error (PRE) measures of association in contingency tables is particularly well done. He uses many examples with real data to excellent advantage in pointing out the differences between and among various measures of association, such as Goodman and Kruskal's Lambda and Tau, two of the more common measures.

Given the enormous advances in methods for the analysis of contingency table data in recent years, many analysts now doing state-of-the-art work have moved to sophisticated multivariate techniques. It behooves the apprentice social scientist, however, to start at the beginning and obtain an understanding of the basis for such techniques before using them. Professor Reynolds provides an efficient vehicle for such a task, and the novice may discover along the way that some interesting insights can be gleaned using relatively simple techniques. Professor Reynolds notes that there are many questions that can be asked about a simple table of data, and the analyst errs when asking only a single—often the most obvious—question to be answered by a simple statistic. Simple tables contain much information, and the type of information summarized by different statistical measures and tests varies, thus suggesting that several tests and measures may be needed to fully comprehend the information contained in each table of interest.

—*John L. Sullivan*
Series Co-editor

ANALYSIS OF NOMINAL DATA

Second Edition

H. T. REYNOLDS
University of Delaware

1. INTRODUCTION

Social scientists face a dilemma. On the one hand, they frequently have to analyze rather crudely measured data. Despite efforts to be as rigorous and precise as their colleagues in the natural sciences, they cannot quantify even some of the most important social and political concepts. Instead, they have to rely on rough and general category labels. On the other hand, they are called on to answer very sophisticated theoretical questions and to make exact predictions. It is as though a physicist were asked to construct a theory of heat but could only classify objects "hot" or "cold," instead of giving them a numerical temperature.

In response to this situation, political scientists, sociologists, statisticians, and others have developed a wide variety of techniques for analyzing categorical or nominal data. Indeed, so many methods have emerged that it is impossible to cover them in one place. Fortunately, however, once a few basic ideas have been grasped, understanding the different procedures becomes easier.

This paper describes methods for analyzing *nominal* variables, variables whose "scores" are category labels such as occupation, sex, or religion. It covers mostly two-way cross-classifications of such variables (e.g., Table 1), although toward the end more complicated situations are briefly discussed. Rather than describing every approach, the paper emphasizes a few fundamentals, including the special difficulties that arise in analyzing nominal data. Concentrating on the basics, one hopes, will make numerical results more meaningful.

TABLE 1
Party Identification and 1980 Presidential Vote

X: Party Identification

Y: Presidential Vote 1980	Strong Democrat	Democrat	Independent Democrat	Independent	Independent Republican	Republican	Strong Republican	Totals
Reagan	11% 21	33% 66	29% 32	64% 54	76% 85	86% 131	92% 103	492
Carter	86% 168	60% 120	45% 49	22% 19	12% 13	5% 7	4% 5	381
Anderson	4% 7	7% 15	26% 28	14% 12	13% 14	10% 15	4% 4	95
Totals	196	201	109	85	112	153	112	968

SOURCE: These data, the 1980 American National Election Study, were made available by the Inter-university Consortium for Political Research through the Center for Political Studies, University of Michigan. The Consortium, of course, is not responsible for any errors or interpretation of these data.

8

Preliminaries

The analysis of nominal data is perhaps best illustrated by an example. The data in Table 1 consist of a sample of 968 adults cross-classified by their political party preference and their 1980 presidential vote.[1] Both variables are of course nominal or categorical because (a) each individual is assigned to one and only one class according to a particular trait or attribute; (b) the category labels are simply names that indicate how groups differ from one another; and (c) the labels say nothing about the magnitude of the differences—indeed, the appearance of the names in any particular order is arbitrary. (We will later rearrange the columns of the table to suit our needs at that time.)

Although the table seems simple enough—tables of this sort are probably familiar to most readers—it can be used to answer a startling variety of questions. One might ask, for example,

—Is there a relationship between party identification and candidate preferences? Alternatively, can knowledge of partisanship be used to predict how people will vote?

—Assuming that there is a relationship in the sample, can one infer that such a relationship holds for the population from which it was drawn?

—Do Democrats behave differently from Republicans? More specifically, which group votes more consistently along party lines and which deviates more?

—Similarly, do weak partisans differ from strong ones?

—Who did Anderson's candidacy hurt most, Carter or Reagan?

—How do independents vote, compared to party identifiers?

—One hears a lot these days about the decline of party; has the relationship weakened over the years?

—Is the relationship between party identification and vote the same among different subgroups in the population, such as college graduates or Southerners?

There are, in short, a host of questions that even a simple cross-classification like Table 1 can address. In fact, it is often self-defeating to limit oneself to a single question to be answered by a simple statistic. There is usually much more information than that in a cross-classification.

Nominal variables. A nominal scale consists of a set of categories representing different realizations of an underlying trait. Ideally, the individuals assigned to a category are homogeneous with respect to the attribute; mixing Democrats and Republicans in the category "Independent" only misinforms us about political behavior. The categories should also be *mutually exclusive* (no one is assigned to more than one category) and *exhaustive.*

The categories of a nominal variable can be arranged in any particular order that suits the needs of the investigator. As we will see, the columns of Table 1 can be rearranged without losing information. By contrast, the categories of an *ordinal* variable do have an implicit order: They measure not only qualitative but quantitative difference. The categories of a variable such as social status (low, medium, and high) cannot be arranged in another order (e.g., medium, low, high) without possibly throwing away valuable information. Hence, nominal and ordinal data are both categorical variables; the difference is that ordinal scales involve an ordering among the groups, whereas nominal variables do not.

Nominal scales may measure truly discrete phenomena such as race or sex, but in most instances they probably represent measurement error of one kind or another because the underlying traits are more or less quantitative. Attitudes, for instance, are not simply a matter of pro or con. Instead, people hold them with varying degrees of conviction. Thus, an attitude might really be considered a continuum running from Strongly Agree through Neutral to Strongly Disagree. Difficulties in measuring attitudes should not obscure this potential richness of information.

In particular, the number and quality of categories is extremely important in making correct inferences. One of the biggest mistakes in social and political research is to lump the respondents into a few categories. Dichotomizing data (putting everyone into one of two groups) for convenience or because everyone else does it is hardly ever justified. Poorly measured data will almost certainly produce misleading substantive conclusions, no matter what statistical technique is used.

Dependent versus independent variables. Most social scientists agree that a person's party identification, usually acquired by adolescence, partly determines his or her political preferences. In this sense, a vote in 1980 would "depend" on partisanship. A variable that depends on, is caused by, or temporally follows another variable is a "dependent variable." And naturally enough, the causal variable is called "independent." A change in the level of an independent variable changes the level of the dependent variable, but changing the dependent variable does not affect the independent variable.

Some of the techniques described here can be applied only when the investigator has a clearly defined dependent variable. Although other procedures make no assumptions about causal dependencies, one should think carefully about causal relationships among variables. It is possible to arbitrarily designate a variable as dependent or independent—this is occasionally necessary and there is nothing in any formula to prevent it—but the results may be wrong.

Of course, these decisions represent assumptions about the data, since it is not possible to prove that one variable causes another.

The number of categories. Since nominal variables are ordinarily organized into two-way or multi-way tables (see Tables 1 and 22, respectively, for examples), it is customary to construct them in such a way that each cell has as many cases as possible. This practice is understandable: Cross-classifications containing numerous zeros do not seem very reliable or impressive. Nevertheless, collapsing or combining categories to increase cell frequencies undoubtedly creates as many problems as it solves. There are two reasons for this.

First, the variation in a nominal variable depends in part on the number of its categories: the greater the number of classes, the greater the variation, other things being equal. Here, "variation" refers to the measured differences among individuals. If all of the cases are in a single category, there is no dispersion or variation among them; if, on the other hand, they are more or less evenly divided among several classes, there is greater variation. Classifying people as Democrats, Republicans, or Independents is simpler than assigning them to more precise groups like "strong Democrat," but there is less variation. And, as in regression analysis, the amount of variation, especially in independent variables, partly affects the magnitude of measures of association.

A second problem is that combining or reducing the number of categories often seriously affects observed interrelationships. Suppose, for example, an investigator has three variables, each having five categories. In order to simplify the presentation of the results, however, this person decides to collapse each variable into two categories. Yet it is likely that the results of most statistical methods based on the dichotomized data will differ from those that would be obtained from the uncollapsed variables, even though the same variables, sample, and techniques are used. One is apt to find associations in the $2 \times 2 \times 2$ table that did not occur in the $5 \times 5 \times 5$ cross-classification.

The lesson, then, is simple: Retain as many categories as possible and do not dichotomize or trichotomize variables without good reason and without attempting to ensure that substantive conclusions have not been affected.

Sampling. All of the techniques presented here apply to simple random samples drawn from a population. The marginal totals in such a sampling scheme are not fixed or predetermined by the investigator. (The column totals at the bottom of Table 1 are the "marginal totals" for party identification.) Referring to Table 1, for instance, 196 strong Democrats appeared in the sample by chance.[2] This number was *not* determined by the research design. Such a sample is sometimes called a "multinominal" sample, and variables generated in this way are sometimes called "responses."

Many of the methods can be applied to other types of samples. In particular, they can be often used when one or more marginal total is fixed before the research begins. For example, an investigator may decide to interview equal numbers of Democrats, Republicans, and Independents. Or he or she may include twice as many partisans as Independents. In either event, the marginal totals associated with party identification have been set ahead of time. Some statisticians call these variables "factors." Of course it is possible to have a mixture of responses and factors, as in a one-factor, one-response cross-classification. Responses are usually considered dependent and factors independent variables. Research designs involving only response variables occur most frequently in sample surveys, while designs involving factors as well as responses arise in laboratory experiments.

Notation. Possibly frightening at first, notation facilitates the presentation and explanation of statistical ideas. The principles are actually quite simple.

In cross-classifying subjects on the basis of two attributes (such as party identification and vote) one creates a "two-dimensional" table. The *dimensionality* of a cross-classification refers to the number of variables. Excluding the totals at the bottom, Table 1 has three rows and seven columns and is referred to as a 3×7 (read "3 by 7") table. If it has I rows and J columns, it is an $I \times J$ table, where I and J take any integer values.

TABLE 2
General Table with I Rows and J Columns

		X (Independent Variable)				
		1	2	j	J	
	1	n_{11}	n_{12} $\cdots$	n_{1j} $\cdots$	n_{1J}	n_{1+}
	2	n_{21}	n_{22} $\cdots$	n_{2j} $\cdots$	n_{2J}	n_{2+}
Y (Dependent Variable)		.	.	.	.	.
	i	n_{i1}	n_{i2} $\cdots$	n_{ij} $\cdots$	n_{iJ}	n_{i+}
		.	.	.	.	.
	I	n_{I1}	n_{I2} $\cdots$	n_{Ij} $\cdots$	n_{IJ}	n_{I+}
Totals		n_{+1}	n_{+2} $\cdots$	n_{+j} $\cdots$	n_{+J}	n

Cross-classifications of more than two variables are called "multi-dimensional" tables. An example might be a table showing the relationship between party identification and vote, controlling for education. (See Table 22.) By convention, the row variable (which is usually the dependent variable) appears first, then the column variable (usually independent), and the "layer" or "control" variable last. It is also convenient to let letters X, Y, and Z denote variables.

Table 2 shows a general I × J table, in which the independent variable is labeled X and the dependent variable is labeled Y. (This convention is used throughout the paper.) The row labels begin with 1 and go to I, which signifies the last category of Y and, similarly, column labels run from 1 to J, the last category of X.

A specific combination of row and column variables is designated by subscripts, the first letter indicating the row category and the second the column category. Small n's represent the cell frequencies of the number of cases in a particular row-column classification. The number of cases in the first row and first column of Table 2 is n_{11}. This is, of course, the number of people who are in the first category of each variable. (In Table 1, n_{11} corresponds to the 21 strong Democrats who voted for Reagan.) Likewise, n_{ij} represents the number of individuals in the ith row and j^{th} column. The subscripts i and j can take on any values from 1 to I and 1 to J respectively.

The totals at the bottom and side ($n_{1+}, n_{2+} \ldots; n_{+1}, n_{+2} \ldots;$ etc.) constitute the *marginal distributions* of X and Y. Looking at the first row, note that n_{1+} means the sum of all the observations in the first row. The plus sign in the subscript indicates that all the entries in the first row have been added over the J columns. By the same token, n_{+1} is the total of all the cases in the first column. In Table 1, for example, $n_{1+} = 21 + 66 + \ldots + 103 = 492$ and $n_{+1} = 21 + 168 + 7 = 196$.

The total number of observations is n. This quantity is found by summing all the n's in the table or by summing the marginal totals of either variable.

Besides working with tables of frequencies, it is often necessary or useful to deal with probabilities. Let $P(Y_iX_j)$ denote the probability that an individual is in the ith class of Y *and* j^{th} class of X in an I × J population cross-classification. $P(Y_1X_1)$ in other words, represents the probability of having characteristic 1 on variable Y (that is, voting for Reagan) *and* characteristic 1 on variable X (that is, being a strong Democrat). To simplify the notation, P_{ij} often replaces $P(Y_iX_j)$.

The layout of probability tables follows the same guidelines as contingency tables. P_{i+}, for instance, means the marginal probability of being in the ith category of Y. It is obtained by adding the probabilities in the ith row:

$$P_{i+} = P_{i1} + P_{i2} + \ldots + P_{iJ}$$

assuming the table has J columns. It gives the probability of being in the ith category of Y, irrespective of X.

Analyzing Nominal Data

The analysis of nominal data normally begins with a two-variable cross-classification. An investigator identifies a pair of variables that on theoretical grounds are assumed to be related in one way or another. Several questions arise in the study of such a relationship:

I. Are the variables statistically independent in the population? This question is usually answered with the familiar chi-square test, described next.

IIa. If the variables are related, what particular combination of categories explains their association? Chapter 2 illustrates two methods for answering this question.

IIb. Again assuming a statistically significant association, what is the magnitude or strength of the relationship? Section 3 presents a few common indices of association, together with a discussion of their properties and interpretations.

III. Finally, is the observed relationship part of a more complex system of interrelationships involving three or more variables? This problem, perhaps the most important of all, leads to multivariate analysis. Although this topic cannot be treated in detail here, some preliminary ideas will be presented in Chapter 4 that may make the more advanced techniques easier to comprehend.

2. CHI SQUARE TEST

The chi square test for independence provides a standard for deciding whether two variables are statistically independent. The test consists of four parts: (1) the null hypothesis (H_o) that the variables are statistically independent; (2) expected frequencies derived under the assumption that the null hypothesis is true; (3) a comparison of these expected values with the corresponding observed frequencies; and (4) a judgment about whether or not the differences between expected and observed frequencies could have arisen by chance.

The chi-square test actually has the same logic as more advanced multivariate procedures: One first states a model or hypothesis. *If* the model is true for a population, then, except for sampling error, one would expect that a sample drawn from it would exhibit certain characteristics. These expected results can be compared with what actually occurs. If the differences between the expected and observed results are small, the conclusion is that they could have arisen by chance, hence, the model is acceptable. On the other hand, if the discrepancies between what is expected and observed are large, one might decide to reject the model in favor of an alternative.

The null hypothesis. The null hypothesis is that in a population the two variables in a cross-classification are statistically independent. More formally, statistical independence holds if:

$$P_{ij} = P_{i+} P_{+j} \quad \text{for all i, j} \tag{1}$$

where P_{ij} is the probability of being in the ith category of Y (the row variable) *and* the jth category of X (the column variable); P_{i+} is the *marginal* probability of being in the ith category of Y; and P_{+j} is the marginal probability of being in the jth category of X. Returning to the example, the probability of having a particular party identification *and* voting for a particular candidate is the product of the corresponding marginal probabilities. If, on the other hand, Democrats, say, are more likely to vote for Carter than are Republicans, then equation 1 would not hold and the variables would be statistically related.

Deriving expected values. The next step is the determination of cell frequencies expected in a sample table if the null hypothesis is true. In a sample of 968, for example, how many cases would we expect to find in the cells of Table 1 if equation 1 were true for the population?

To make these calculations, consider this reasoning: If there is no association between party and vote, the proportion of Reagan voters should be the same in all seven categories of partisanship. If the proportion of Reagan voters overall is .5 and statistical independence holds, then 50 percent of strong Democrats, 50 percent of Democrats, and so on down the line should have voted for Reagan. The true proportion of Reagan voters is unknown, but it can be estimated from the sample as

$$\frac{492}{968} = .508$$

Consequently, under the hypothesis of independence, one would expect 50.8 percent of the 196 strong Democrats to have voted for Reagan. In other words, the first cell of the table should contain

$$\frac{492}{968} \times 196 = .508 \times 196 = 99.62$$

cases. That is, there should be about 99 to 100 cases in the first cell if the null hypothesis is true.

Other expected frequencies are found in a similar way. Assuming independence, one would expect about 50.8 percent of, say, the 112 weak Republicans to have voted for Reagan, or

$$\frac{492}{968} \times 112 = 56.93$$

These considerations lead to the following formula for estimating the expected frequencies under the null hypothesis:

$$\hat{n}_{ij} = \frac{(n_{i+})(n_{+j})}{n} \qquad [2]$$

where $\hat{n}_{ij}$ is the estimated expected frequency and n_{i+} and n_{+j} are the ith and j^{th} row and column marginal frequency, respectively.

Applying formula 2 to each of the row and column totals gives the expected frequencies shown in Table 3.

Comparing observed and expected frequencies. As one can tell from even a cursory glance at Table 3, there are large discrepancies between observed and expected values. In the first cell, for example (i.e., strong Democrats who voted for Reagan), the observed frequency is 21, whereas, if the null hypothesis were true, we would expect to find about 99 or 100 cases. The differences between observed and expected values are just as large in many other cells of the table, leading one to wonder if the null hypothesis is tenable. After all, if partisanship and vote were statistically independent in the population, should not there be more observations in the first cell?

Still, the agreement is relatively close in other parts (among weak Democrats who voted for Carter, for example), suggesting the need for a systematic method for comparing observed and expected frequencies among all cells of the table. A useful statistic for making this overall comparison is the *goodness-of-fit chi square*:

$$X^2 = \sum_i \sum_j \frac{(n_{ij} - \hat{n}_{ij})^2}{\hat{n}_{ij}} = \sum_i \sum_j \frac{n_{ij}^2}{\hat{n}_{ij}} - n \qquad [3]$$

TABLE 3
Observed and Expected Frequencies and Components
of Chi Square for Table 1

			Party Identification				
	SD	D	ID	I	IR	R	SR
	21	66	32	54	85	131	103
Reagan	99.62	102.16	55.40	43.20	56.93	77.76	56.93
	62.05	12.80	9.88	2.70	13.85	36.44	37.29
	168	120	49	19	13	7	5
Carter	77.14	79.11	42.90	33.46	44.08	60.22	44.08
	107.00	21.13	.87	6.25	21.92	47.03	34.65
	7	15	28	12	14	15	4
Anderson	19.24	19.73	10.70	8.34	10.99	15.01	10.99
	7.78	1.13	27.99	1.60	.82	0.0	4.45

KEY: SD = strong Democrat; D = Democrat; ID = Independent, leaning Democrat; I = Independent; IR = Independent, leaning Republican; R = Republican; SR = strong Republican

NOTE: Row 1 contains raw or original frequencies; row 2 expected frequencies, $\hat{n}_{ij}$; and row 3 the components of the chi-square statistic, e_{ij}.

where n_{ij} and $\hat{n}_{ij}$ are the observed and expected frequencies in the ij^{th} cell; n is the table total; and the summation is over all I rows and J columns. For the data in Table 3, goodness-of-fit chi square is

$$X^2 = \left[\frac{(21)^2}{99.62} + \frac{(66)^2}{102.16} + \dots + \frac{(4)^2}{10.99} \right] - 968 = 457.63$$

An alternative statistic, which for large enough samples will lead to the same conclusion as X^2, is the *likelihood ratio chi square*. Since it is useful for dissecting a cross-classification, one ought to note its formula:

$$L^2 = 2 \sum_i \sum_j n_{ij} \left[\log\left(\frac{n_{ij}}{\hat{n}_{ij}}\right) \right] \qquad [4]$$

where n_{ij} and $\hat{n}_{ij}$ are defined as before and log denotes the natural logarithm. Returning once again on Table 3, we have

$$L^2 = 2\left\{21\left[\log\left(\frac{21}{99.62}\right)\right] + 66\left[\log\left(\frac{66}{102.16}\right)\right] + \ldots + 4\left[\log\left(\frac{4}{10.99}\right)\right]\right\}$$

$$= 2\left\{21\,[-1.5568] + 66\,[-.4369] + \ldots + 4\,[-1.0107]\right\}$$

$$\cong 501.12$$

Although X^2 and L^2 are not equal (as is usually the case), they both lead to the same conclusion about the variables.

Decision. These observed values of X^2 and L^2 are quite large, but since they are computed from a sample one can fairly ask if they could have arisen by chance from a population in which the variables were really independent. Fortunately, if certain conditions are satisfied, one can determine the probability of observing a chi square as large or larger than the one actually observed. The conditions or assumptions are relatively straightforward:

(1) a random sample of n observations;

(2) mutually exclusive and exhaustive categories such that each observation is placed in one and only one cell;

(3) most (e.g., more than 80 percent) of the estimated expected frequencies are larger than 5.

If these conditions hold, one can compare the observed chi square statistic (either X^2 or L^2) with a so-called critical chi square, $x^2\alpha$ to decide whether or not the observed value arose by chance. If X^2 (or L^2) is less than the critical value, one accepts the null hypothesis (H_o) that the variables are independent; if, on the other hand, X^2 (or L^2) equals or exceeds $x^2\alpha$, then the null hypothesis is rejected, and one would conclude that the variables probably are related. The decision, then, rests on a comparison of an observed chi square with a standard that has been determined beforehand—that is, before the data have been examined.

All that remains is the selection of the critical value, $\chi^2\alpha$. The choice depends on two additional considerations: the *degrees of freedom* in the table and the desired level of significance. The degrees of freedom associated with any two-way cross-classfication table are easily computed from the following formula:

$$df = (I - 1)(J - 1) \qquad [5]$$

where I and J are the total number of rows and columns, respectively. For Table 1, df = (3 – 1) (7 – 1) = 12.

The desired level of significance, α, which should be established prior to seeing the data, determines the specific critical value. In order to find it one refers to a tabulated distribution of chi square such as ones found in Blalock (1979) or Reynolds (1977a). These tabulations give critical values for different levels of probability and degrees of freedom. (The degrees of freedom are read down the side of the table, while levels of significance are read across the top.) Given 12 degrees of freedom, for example, the critical values at the .05, .01, and .001 levels are, respectively, 21.026, 26.217, and 32.909. Since both X^2 and L^2 clearly exceed these values, one would reject the null hypothesis of independence between voting and party preference.

There are times when a chi square test is not appropriate. Given a 2×2 table where n is small (say less than 20), Fisher's test, which gives exact rather than approximate probabilities for the observed table, should be used. (See Blalock, 1979: 292-297.) Furthermore, the X^2 statistic only approximates the theoretical chi square. In order to ensure the reasonableness of the approximation, many statisticians recommend computing X^2 only if most *expected* cell frequencies (say 80 percent) are greater than or equal to five. Although there is considerable controversy about the topic, this seems to be a safe rule that can be met in most situations where n is relatively large.

Interpreting the Chi Square Test

When we are told one baseball team defeated another, we usually want to know much more, such as the score, the winning and losing pitchers, and so forth. Likewise, a "significant" chi square tells us that two variables are probably related in the population, but by itself, it reveals little else. What is worse, if used uncritically, the test can mislead as much as it informs.

As with any statistical test, the sample size affects its magnitude. Large samples frequently produce significant chi squares even if the variables are weakly related. Consequently, we may decide to reject the hypothesis of independence when in fact the variables have little or no practical or substantively interesting relationship.

Following an example given in Blalock (1979), the data in Table 4 illustrate the point. The nonsignificant goodness-of-fit chi square in **Table 4a reflects the trivial relationship between X and Y. The distributions in columns 1 and 2 are the same and differ only slightly from the third. Hence, scores on X do not reveal much about Y values and vice versa.**

Furthermore, this conclusion is not changed by multiplying every frequency by 10, as in Table 4b. Since the *proportions* remain exactly the same, the X-Y relationship has not changed at all. Yet, the chi square is now significant. If we looked at only this value without paying attention to the distribution of the cases in the table, we would conclude X and Y are strongly related. These tables illustrate a well-known principle: If all the frequencies in a cross-classification are multiplied by a constant, K, the magnitude of chi square increases K times. Therefore, in order to interpret meaningfully chi square, we have to look at more than its numerical value. Assessing the "practical" or "theoretical" significence of a chi square requires that we measure the strength of the relationship.

TABLE 4
Sample Size Affects the Numerical Magnitude of Chi Square

(Hypothetical Data)

		a					b		
		X		Totals			X		Totals
	30	30	30	90		300	300	300	900
Y	30	30	36	96	Y	300	300	360	960
	40	40	34	114		400	400	340	1140
Totals	100	100	100	300	Totals	1000	1000	1000	3000

$X^2 = 1.38$ with 4 df $X^2 = 13.82$ with 4 df

$\phi^2 = .005$ $\phi^2 = .005$

NOTE: Table entries are frequencies.

The conventional chi square test presents another problem of interpretation. As normally applied, it is not directed at any specific alternative hypothesis. An alternative hypothesis might be, for example, that the variables are related in a particular way. Since the X^2 measures only departures of expected from observed values (i.e., $n_{ij} - \hat{n}_{ij}$), it does not call attention to any "pattern of deviations . . . that may hold if the null hypothesis of independence is false" (Cochran, 1954: 417). If we can specify the form of a possible relationship *before* making the test, we can sometimes improve the efficiency and sensitivity of the procedure.

Thus, once an investigator finds a significant chi square, the work has just begun. Since one often knows ahead of time that two variables are statistically related, not much information is gained from a significance test alone. Instead, it is possible and desirable to further analyze data by specifying more precisely how the variables are related. There are at least two quick and simple methods for doing so: examining the components of the chi square statistic and partitioning the original table into subtables, each pertaining to a particular subhypothesis or question.

Components of the chi square. An easy yet effective procedure is to examine each component of the chi square statistic:

$$e_{ij} = \frac{(n_{ij} - \hat{n}_{ij})^2}{n_{ij}} \qquad [6]$$

These numbers, which are analogous to residuals in regression analysis, indicate which cells contribute most to the chi square and, hence, which categories of variables are most closely related. Table 3 shows, for example, that the cells pertaining to strong partisans (both Democrat and Republican) in the Carter and Reagan rows contribute the most to the relationship, categories involving Independents and Anderson the least. What makes the relationship so strong, apparently, is the differential behavior of Democrats and Republicans toward the main candidates. Over half of the total X^2 is due to just four cells: (1,1), (2,1), (1,7), (2,7). Although this finding makes intuititive sense, in a more complex table such an analysis might reveal less obvious insights.

Some authors recommend computing standardized residuals ($r_{ij} = ((n_{ij} - \hat{n}_{ij})/\sqrt{\hat{n}_{ij}})$ or adjusted residuals, which can be interpreted as standard normal deviates. These statistics make the detection of extreme cases or outliers particularly easy. The interest reader should consult Haberman (1973) and Reynolds (1977) for further details.

Partitioning chi square. Partitioning offers another simple method for more precisely analyzing a cross-classification. It is especially useful because we can test various subhypotheses. In other words, a complex table such as Table 1 may contain a wealth of information that the overall chi square masks.

The basic idea is simple: A table with $(I - 1)(J - 1)$ degrees of freedom can be divided into various subtables. Each subtable is analyzed as though it were a separate cross-classification. Therefore, one can calculate a series of sub-chi squares, each with appropriate degrees of freedom. Many of these subtables can be further divided into still more subtables, each having a chi square and degrees of freedom. In particular, an original $I \times J$ table can ultimately be partitioned into $(I - 1)(J - 1)$ 2×2 subtables, each with one degree of freedom.

These subtables are formed from the original tables in such a way that specific questions and hypotheses are addressed. Table 1, for instance, will be partitioned so that we can assess, among other things, the impact of strong partisanship. The total chi square is thus divided into parts, each pertaining to a separate table and hypothesis. If we calculate L^2 instead of X^2, the sum of the component chi squares will equal the total, as will the degrees of freedom.

Of course, the subtables cannot be formed arbitrarily. Certain rules have to be followed to ensure that the tables are independent and do not contain redundant information.

Professor Gudmund Iversen (1979) provides a simple algorithm for finding suitable subtables. He first divides the frequencies in a contingency table into two types:

A frequencies are either cell frequencies (n_{ij}) or the table total (n). In Table 1, for example, A frequencies are (21, 66, 32 . . . 14, 15, 4) and the table total, 968.

B frequencies are row and column marginal totals, n_{i+} and n_{+j}. In Table 1, B frequencies are the marginal totals (196, 201 . . . 112) and (492, 381, 95).

Iversen's rules for partitioning rely on these two types of frequencies:

Rule 1: Each frequency (whether *A* or *B*) in the original table must appear once and only once as a frequency of the *same* type in one of the subtables. (For example, n must appear in one of the

24

TABLE 5
Table of Party Identification by Vote in 1980 Rearranged

| | | Party Identification | | | | | | |
| | | Partisans | | | | Independents | | |
		SD	SR	D	R	ID	IR	I
	Reagan	21	103	66	131	32	85	54
1980 Vote	Carter	168	5	120	7	49	13	19
	Anderson	7	4	15	15	28	14	12
	Totals	196	112	201	153	109	112	85

KEY: (See Table 3).

NOTE: $L^2 = 501.1230$; 12 df.

subtotals as a total; similarly, each n_{i+} and n_{+j} must appear as a row and column total, respectively.)

Rule 2: Each frequency that is in a component or subtable but not in the original table must appear as a frequency of the other type in another subtable. In other words, subtables contain some frequencies, like marginal totals, that do not appear in the original table; these "new" frequencies have to be used again, so to speak, but as frequencies of a different type.

As with many statistical methods, it is perhaps easier to illustrate the procedure than to describe it verbally. Table 5 shows a simple rearrangement of Table 1 in which partisans (both Democratic and Republican) have been grouped together. This new arrangement is of course legitimate since category labels are arbitrary. Rewriting the columns in this fashion makes it easier to explore various subhypotheses.

Tables 6, 7, and 8 show one of many possible partitions of Table 5. Formed from the first four columns of Table 5, Table 6, for example, explores the voting behavior of partisans. As one can see, $L^2 = 398.240$, suggesting that about 80 percent ($398.240/501.123 = 79.47$) of the

TABLE 6
Voting Behavior Among Partisans

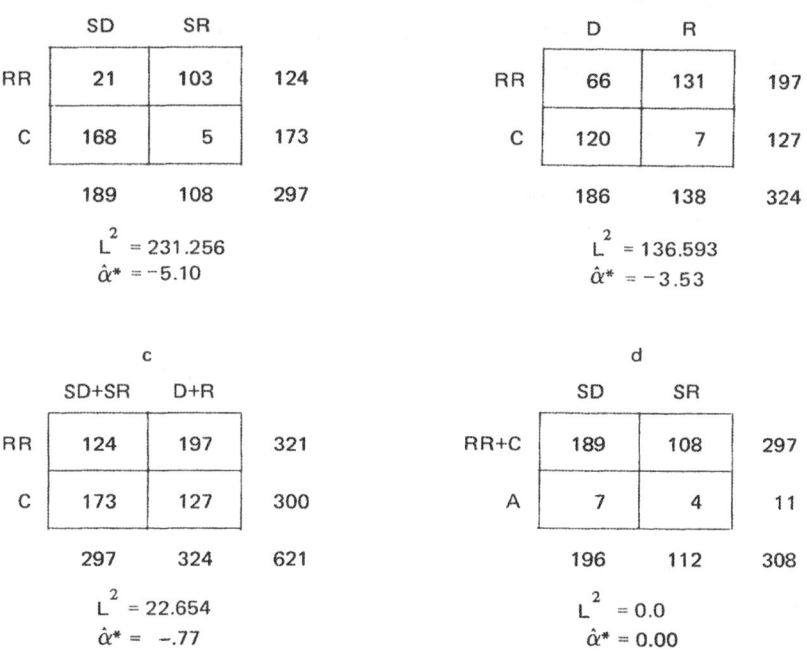

Partition A

	SD	SR	D	R	
RR	21	103	66	131	321
C	168	5	120	7	300
A	7	4	15	15	41
	196	112	201	153	662

$$L^2 = 398.240; 6 \text{ df}$$

Further Partitioning of A

a

	SD	SR	
RR	21	103	124
C	168	5	173
	189	108	297

$$L^2 = 231.256$$
$$\hat{\alpha}^* = -5.10$$

b

	D	R	
RR	66	131	197
C	120	7	127
	186	138	324

$$L^2 = 136.593$$
$$\hat{\alpha}^* = -3.53$$

c

	SD+SR	D+R	
RR	124	197	321
C	173	127	300
	297	324	621

$$L^2 = 22.654$$
$$\hat{\alpha}^* = -.77$$

d

	SD	SR	
RR+C	189	108	297
A	7	4	11
	196	112	308

$$L^2 = 0.0$$
$$\hat{\alpha}^* = 0.00$$

(continued)

Table 6 (Continued)

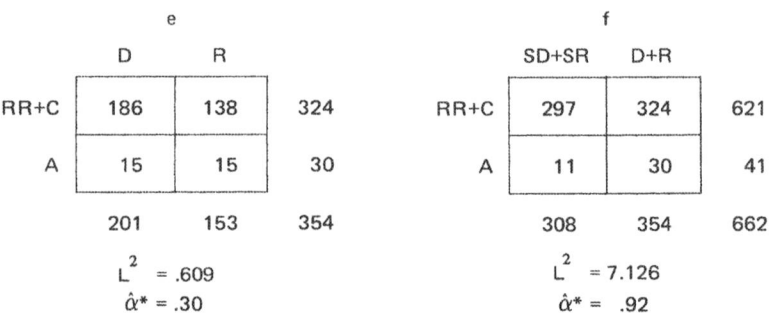

	e				f		
	D	R			SD+SR	D+R	
RR+C	186	138	324	RR+C	297	324	621
A	15	15	30	A	11	30	41
	201	153	354		308	354	662

$$L^2 = .609$$
$$\hat{\alpha}^* = .30$$

$$L^2 = 7.126$$
$$\hat{\alpha}^* = .92$$

KEY: RR = Ronald Reagan; C = Carter; A = Anderson (also see Table 3).

original chi square is due to the differential behavior of Democrats and Republicans.

In addition, Table 6, with six degrees of freedom, has itself been partitioned into six more tables, each with one degree of freedom. Also note that all of these partitions follow the two rules above. Each of the frequencies in Table 6, for instance, appear in one of the component tables (Tables 6a to 6f) as a frequency of the same type (Rule 1) and each "new" frequency (a frequency that was not in Table 6 to start with) appears twice, once as an A frequency and once as a B frequency. (The row marginal total of 124 in Table 6a—a B type frequency—appears as an A type frequency in the first cell of Table 6c). The reader can quickly verify that both rules have been followed in making this partition.

The decomposition of Table 6 leads to further insights. An enormous part of the relationship in Table 6 (and indeed in the original table as well) is due to differences between strong Democrats and strong Republicans in their votes for Reagan and Carter; similarly, the differences between plain Democrats and Republicans are large (Table 6b). Together these two chi squares account for the lion's share of the relationship in Table 6. Anderson's candidacy had relatively little effect on partisans because they mostly stayed with their party's candidate; the two chi squares total only .609.

Table 7, formed from Table 4 according to the two rules, shows the political behavior of Independents and weak partisans. It and the

TABLE 7
Voting Behavior Among Independents

Partition B

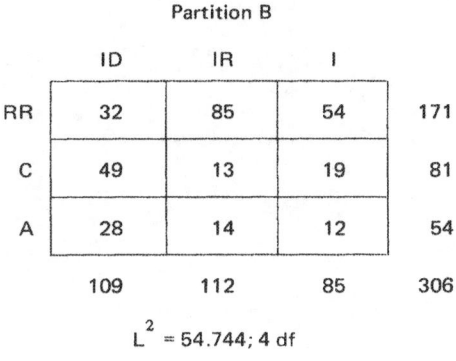

	ID	IR	I	
RR	32	85	54	171
C	49	13	19	81
A	28	14	12	54
	109	112	85	306

$L^2 = 54.744$; 4 df

Further Partitioning of B

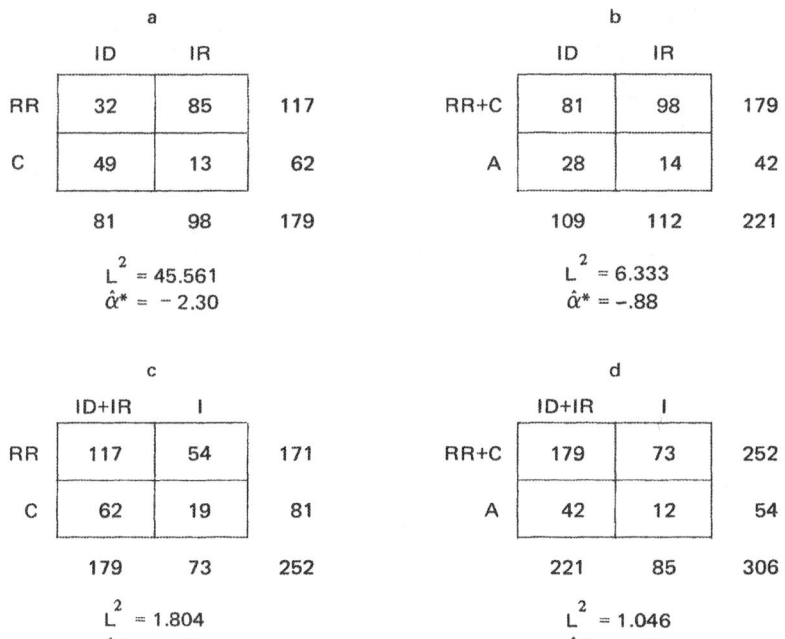

a

	ID	IR	
RR	32	85	117
C	49	13	62
	81	98	179

$L^2 = 45.561$
$\hat{\alpha}^* = -2.30$

b

	ID	IR	
RR+C	81	98	179
A	28	14	42
	109	112	221

$L^2 = 6.333$
$\hat{\alpha}^* = -.88$

c

	ID+IR	I	
RR	117	54	171
C	62	19	81
	179	73	252

$L^2 = 1.804$
$\hat{\alpha}^* = -.41$

d

	ID+IR	I	
RR+C	179	73	252
A	42	12	54
	221	85	306

$L^2 = 1.046$
$\hat{\alpha}^* = -.36$

TABLE 8
Voting Behavior of Partisans Versus Independents

Partition C

	Partisans (SD+SR+D+R)	Independents (ID+IR+I)	
RR	321	171	492
C	300	81	381
A	41	54	95
	662	306	968

$$L^2 = 48.138; 2df$$

Further Partitioning of C

a

	Partisans (SD+SR+D+R)	Independents (ID+IR+I)	
RR	321	171	492
C	300	81	381
	621	252	873

$$L^2 = 19.422$$
$$\hat{\alpha}^* = -.68$$

b

	Partisans (SD+SR+D+R)	Independents (ID+IR+I)	
RR+C	621	252	873
A	41	54	95
	662	306	968

$$L^2 = 28.716$$
$$\hat{\alpha}^* = 1.18$$

KEY: (See Tables 3 and 6).

subsequent partition (Tables 7a to 7d) illustrate that even among weak partisans, people tend to follow their party identification, but Anderson did better among weak Democrats than among Republicans.

Table 8 finally completes the partitioning by comparing partisans with Independents. Clearly the relationships are smaller here than in Table 6, but they illustrate that partisans vote a more consistent party line (see Table 8b). These findings suggest that party identification is still an important determinant (or at least codeterminant) of voting. If partisanship has become less influential, as many claim, the decline is probably due to fewer people identifying with a party than to a weakening of the relationship. As an interesting exercise, the reader

TABLE 9
Summary of the Partition of Table 5

Source	L^2	df	Sig.	Log Odds $(\hat{\alpha}*)$
Partisans (Table 6)				
a SD/SR v. Reagan/Carter	231.256	1	.001	−5.10
b D/R v. Reagan/Carter	136.593	1	.001	−3.53
c SD+SR/D+R v. Reagan/Carter	22.654	1	.001	−.77
d SD/SR v. Major/Anderson	0.000	1	NS	0.0
e D/R v. Major/Anderson	.609	1	NS	.30
f SD+SR/D+R v. Major/Anderson	7.126	1	.01	.92
	398.240	6		
Independents (Table 7)				
a ID/IR v. Reagan/Carter	45.561	1	.001	−2.30
b ID/IR v. Major/Anderson	6.333	1	.001	−.88
c ID+IR/I v. Reagan/Carter	1.804	1	NS	−.41
d ID+IR/I v. Major/Anderson	1.046	1	NS	−.36
	54.744	4		
Partisans v. Independents (Table 8)				
a Partisans and Independents v. Reagan/Carter	19.422	1	.001	−.68
b Partisans and Independents v. Major/Anderson	28.716	1	.001	−1.18
Total	48.138	2		
Total	501.123	12		

KEY: (See Table 3).

could compare the strength of the relationship across time, using the partitioning described above.

Table 9 summarizes the partitioning of Table 4. It shows once again that out of a total chi square of 501.123, the vast majority (231.256 + 136.593 + 45.561 = 413.410, or 82 percent) is due to 12 out of 21 cells: those pertaining to strong and weak partisans and Reagan and Carter votes. Although these findings are consistent with existing theory and common sense, partitioning a complex cross-tabulations may lead to new and unexpected results. It should also be emphasized here that the partitioning given in Tables 6, 7, and 8 is not the only one possible. Others based on different substantive questions and lines of reasoning might be worth pursuing. The only constraints are the two rules described by Iversen. For further information and examples, the reader should consult Reynolds (1977a), Iversen (1979), and Goodman (1969).

3. MEASURES OF ASSOCIATION

Introduction

Since the usual goodness-of-fit chi square test says only whether two variables may or may not be statistically independent, and by itself it does not give the strength or form of the relationship, one usually calculates a *measure of association* as the next step in the analysis.

A measure of association is a numerical index summarizing the strength or degree of relationship in a two-dimensional cross-classification. Depending on its meaning, it may also reveal how the variables are related to each other. There are literally dozens of measures from which to choose. Three considerations guide the choice of a measure: whether it is symmetric or asymmetric, its interpretation, and its sensitivity to confounding influences.

Symmetric versus asymmetric measures. If a theory or common sense indicate that one variable causes another, then it is usually necessary to predict values of the dependent variable from knowledge of the causal or independent variable. In this case the most appropriate measure would be *asymmetric*. The calculation and interpretation of asymmetric measures depend on which variable is considered dependent. There are thus two versions—one when X is considered independent, the other when it is considered dependent—and the versions do not normally equal each other. With a *symmetric* index, by contrast, either variable can be considered dependent.

The interpretation of measures of association. Since a measure of association—a single number—supposedly summarizes the information in a table, it should have a clear interpretation. The numerical value of most measures lie between 0 and 1.0—zero if the variables are completely unrelated according to some definition of a nil relationship, and 1.0 if the variables are perfectly associated, again according to some criterion of "perfect." The meaning of intermediate values depends on how the measure is operationally defined.

Nil associations. To say that the association between two variables is "nil" usually—but not always—implies that they are statistically independent. As noted above, statistical independence means that the probability of the joint occurrence of two events—that is, for example, the probability of being *both* a Democrat *and* a Reagan voter—equals the product of the probabilities of their separate occurrences. If independence holds for a population table, most measures of association are defined so that they will be zero. For samples drawn from that population, the measures will equal zero subject to sampling error. Thus, values close to zero typically indicate a weak to nil relationship.

A few measures of association are zero even in the absence of statistical independence. Lambda, an index described later, frequently equals zero when the marginal totals are highly skewed—that is, most cases fall into one category—but the variables are not independent. Many social scientists view this property as an inherent weakness in the measure. On the other hand, one can define "nil" to mean something besides independence (Weisberg, 1974). Such definitions are rare, however, and throughout this paper statistical independence is the definition of nil association.

Perfect associations. Having only rough classifications, an investigator may not worry about defining or measuring perfect association. After all, classified variables often represent measurement error where individuals are lumped together into categories out of convenience or necessity. Were more refined divisions available, the observed pattern of relationship might be different. Hence, defining perfect association for nominal data might be premature; if nothing else, it assumes the meaningfulness of the observed classifications. Nevertheless, it is common to conceptualize perfect association even for categorical variables. There are at least three different ways of doing so.

(1) *Strict Perfect Association.* Under strict perfect association, each value of one variable is uniquely associated with a value of the other. Table 10a presents an example. Note that for this condition to hold, the numbers of categories of Y and X must be equal. Here knowledge of a person's X category implies perfect prediction of his or her score on Y. Under this relationship, "normed" measures of association equal 1.0.

(2) *Implicit Perfect Association.* Since one variable frequently has more classes than another (that is, the rows and columns are not

TABLE 10
Models of Perfect Association for Nominal Variables

	a					b			
	Strict Perfect Association					Implicit Perfect Association			
	X					X			
	50	0	0			0	50	0	50
Y	0	0	50		Y	50	0	0	0
	0	50	0			0	0	50	0
Totals	50	50	50		Totals	50	50	50	50

	c		
	Weak Perfect Association		
	X		
	50	0	0
Y	50	0	0
	50	50	50
Totals	150	50	50

NOTE: Table entries are the number of cases.

equal), the definition of perfect association has to be modified. In Table 10b, for example, a column category completely specifies a row category. In other words, the members of a column classification are as homogeneous as possible with respect to Y in the sense that there is only one nonzero row entry per column. Different X categories are generally associated with different Y categories, but since the classes on X outnumber those on Y, the association is not unique.

Not every measure of association achieves its maximum values here. Hence, if one feels this type of table represents a perfect association, the choice should be a measure that does attain its maximum.

(3) *Weak Perfect Association*. Table 10c illustrates a relationship that a few social scientists consider perfect. Here, the categories of, say, X are as homogeneous as possible with respect to Y, given the differences in the variables' marginal totals.

Most measures do not attain their maxima for relationships of this sort. In Table 10c the problem lies in the first column and last row, which contain the preponderance of cases. Knowing an individual's score on one of these classes does not help predict his or her classification on the other variable. Furthermore, if more refined classifications could be obtained, one still has no assurance that the association would be perfect in any of the above senses. Nonetheless, if it seems that these data exhibit perfect association, one should choose an appropriate measure. Unfortunately, the only strictly nominal measures that attain their maximums in this case are applicable solely for 2×2 tables. (If the variables are both ordinal—that is, they have ordered classes—then other indices are available.)

Intermediate values. Since most indices are normed so that they have intelligible lower and upper bounds, one has some basis for choosing among them. He or she simply selects the one based on the definitions of nil and perfect relationships most in agreement with research needs.

The rub lies in interpreting intermediate values. Suppose that an investigator wants statistical independence and strict association to define nil and perfect association. The choice of measures is narrowed somewhat but the investigator still has to make sense of values lying between 0 and 1.0. Suppose the value of an index is .45. What does he or she conclude about the strength and form of the relationship?

The answer naturally turns on the measure's operational definition. Some, like measures based on chi square, do not have intuitively appealing interpretations. Others, proportional-reeducation-in-error indices, for example, are more easily understood but depend on looking at a cross-classification in a particular way. Thus, each measure has to be examined separately in order to grasp its underlying logic and meaning.

Confounding factors. A problem common to all indices is that extraneous factors frequently confuse their interpretation. It is well known that the number of cases affects the magnitude of the chi square statistic: the greater the sample size, the larger the value of chi square, other things being equal. Almost all measures of association eliminate the effects of sample sizes—this is one reason for calculating them in the first place—but similar types of factors can influence their numerical values. The two most common problems are skewed marginal distributions and unequal numbers of rows and columns.

Skewed marginal distributions. Marginal distributions affect the numerical values of many measures of association. In the first panel of Table 11a, most of the cases fall in the middle column, whereas in Table 11b, they are more evenly distributed among the categories of X. Clearly, the second data set involves more variation on X. But also notice that the column percentages (or relative frequencies) are the same in both tables.

In spite of the equivalence in the relationships (at least as measured by percentages), many measures do not give the same value for both tables. A researcher who computes an index for the second set of data might report a strong relationship, while someone analyzing the first table might find a much weaker association, even though they both use the same statistic.

TABLE 11
Categorical Data with Different Marginal Distributions
but the Same Inherent Relationship

		a					b			
		X		Totals			X		Totals	
	60%	20%	10%			60%	20%	10%		
	(60)	(200)	(10)	270		(180)	(120)	(30)	330	
Y	30	60	30		Y	30	60	30		
	(30)	(600)	(30)	660		(90)	(360)	(90)	540	
	10	20	60			10	20	60		
	(10)	(200)	(60)	270		(30)	(120)	(180)	330	
Totals	100%	100%	100%		Totals	100%	100%	100%		
	(100)	(1000)	(100)	1200		(300)	(600)	(300)	1200	

NOTE: Numbers in parentheses are the number of cases.

Only a few indices are impervious to marginal distributions. As a result, an investigator has to pay particular attention to marginal totals. When one or both variables are highly skewed, it should be decided whether or not the relative absence of variation is substantively meaningful. If one wants to know what more even distributions would produce, the observed data can be adjusted (using a method described shortly) or a less sensitive measure selected. On the other hand, the lack of variation may itself be theoretically important, and in that case one would want to preserve the original marginal distributions. The point is simply that it is necessary to be aware of the possible confounding effects of marginal distributions.

Nonsquare tables. The number of rows and columns frequently are not equal. Tabular asymmetry usually occurs by happenstance and ought not to disturb the inherent relationship between two variables. Yet a surprisingly large number of measures are affected by it.

The difficulty arises because some measures cannot attain their maximums if the rows do not equal the columns. Suppose interest lies in the hypothesis that two variables are "implicitly" perfectly correlated. In order to test this proposition, one requires an appropriate measure that can attain its maximum in nonsquare tables.

Measures of Association for 2×2 Tables

Perhaps the best-known and most extensively studied type of cross-classification is the 2×2 table. Formulas and calculations are usually much simpler in the 2×2 case than in larger tables. In addition to their simplicity, they have interesting and useful properties. Many seemingly different measures of associations equal each other in these tables and, as will become apparent, many concepts applicable in the dichotomous case readily generalize to higher dimensional tables.

Nevertheless, there is little advantage in "collapsing" or reducing a larger array into a 2×2 table. Collapsed data frequently introduce distortions. What might be a weak relationship in and I X J table where I and J are both greater than 2 could turn out to be a large association if the variables have been dichotomized. The observed relationship would then be more an artifact of one's measurement than a reflection of the true state of affairs. Of course, if the variables really have only two categories or no other measurement is available, a researcher has little choice. But the widespread practice of uncritically dichotomizing variables risks throwing away valuable information and producing misleading results (Reynolds, 1977b).

Percentages. Regardless of the size of a table, one of the easiest ways to measure a relationship between two variables, especially if one is clearly dependent, is to calculate percentages. After all, one wants to compare how people in different categories of one variable behave with respect to the classes of another. If the distribution of responses changes from one category to another, there is evidence for a relationship.

Table 1 illustrates the point. Comparing Democrats, Independents, and Republicans, one sees a steady decline in the percentage or proportion of support for Reagan. Thus, percentages permit one to detect patterns of departure from independence. That is, besides seeing that two variables are not independent (as chi square test would indicate), one can see *how* the variables are related. If Democrats tended to vote for Reagan instead of Carter, and Republicans for Carter instead of Reagan, the variables would still be related, but in a different fashion. Hence, percentages help one distinguish different forms of association.

Percentages are particularly useful in 2 × 2 tables. A difference in percentages (or proportions) can be interpreted as a regression coefficient between two dichotomous variables. (A regression coefficient gives the magnitude of a change in Y, considered the dependent variable, for a unit change in the independent variable.) Consider this very simple 2 × 2 table of proportions, where raw frequencies are in parentheses:

X

	.9	.4
Y	(45)	(20)
	.1	.6
	(5)	(30)
	1.0	1.0
	(50)	(50)

Here the difference in proportions (with respect to the first row) between the two columns of X is .5. The same quantity would be obtained if the categories of X and Y were coded 0 and 1 and the data substituted into familiar regression formulas. A change in one unit of X (from 0 to 1)

37

TABLE 12
**Relationship Between Party Identification and Candidate
Preference When the Variables Are Dichotomized**

Y: Presidential Vote, 1982	X: Party Identification		
	Democrats	Republicans	Total
Reagan	87	234	321
Carter	288	12	300
Totals	375	246	621

NOTE: Table entries are the number of cases. Independents and Anderson voters have been excluded. "Democrats" include both strong and regular Democrats. Likewise, "Republicans" include both strong and regular Republicans.

produces a change of .5 in Y. Given the range of possible values (0 to 1), this result indicates a substantial relationship.

Thus, a difference in percentages or proportions has a clear interpretation as a measure of association in a 2 × 2 table. Furthermore, by virtue of its definition, it is not sensitive to imbalances in the marginal distribution of X.

Being asymmetric, the calculation and meaning of a difference in percentages assume that the dependent and independent variables have been unambiguously specified. And, of course, one needs a sufficiently large number of cases in each category of X (usually 20 or more) in order to obtain a reasonable estimate of its value, a remark that applies to the computation of percentages in general.

Cross-product ratio. The cross-product ratio, often called the odds ratio, is surprisingly little known in the social sciences—surprisingly because it actually underlies two popular measures of association and has several useful properties. A thoroughly researched statistic, the ratio also provides a very helpful heuristic device for understanding log-linear analysis, a categorical multivariate technique (see Knoke and Burke, 1980).

A simplified version of the voting data (see Table 12) illustrates its meaning and computation. (An investigator would not normally analyze these data in this manner; it is done here for convenience.) The new table shows the relationship between two dichotomies: party identification (Democrats and Republicans) and vote. The marginal totals are considered fixed.

Obviously the variables are related; but the question is, how strongly? One answer is given by comparing the *odds* of voting for Reagan. For Democrats these odds are 87 to 288, or about .3 to 1. (More precisely, $87/288 = .302$.) Now compare these odds to the odds of voting for Reagan among Republicans. If partisanship were unrelated to ideology, the odds of being a Reagan voter should be the same for both Democrats and Republicans. If, on the other hand, party identification affects people's preferences, then the odds among Republicans will be different. As it turns out, the odds of voting for Reagan among Republicans is considerably greater than 1, namely $234/12 = 19.500$.

Although the odds for the two groups obviously differ, it is useful to compare them more explicitly by calculating their ratio:

$$\hat{\alpha} = \frac{\dfrac{n_{11}}{n_{21}}}{\dfrac{n_{12}}{n_{22}}} = \frac{n_{11} n_{22}}{n_{21} n_{12}} = \frac{\dfrac{87}{288}}{\dfrac{234}{12}} = \frac{.302}{19.500} = .02$$

The ratio of the odds (denoted $\hat{\alpha}$) has a simple interpretation. If the odds are the same in both categories of party identification, their ratio will equal 1.0. As an example, the ratio for the hypothetical data

	X	
	45	90
Y		
	15	30

is $45/15/90/30 = 3/3 = 1.0$. Hence, 1.0 indicates no relationship. This definition has intuitive appeal because if the odds of being for Reagan are the same in both classes of party identification, it (partisanship) provides very little insight into how people will vote.

Departures in either direction from 1.0 suggest association: the greater the departure, the stronger the relationship. The small value of $\hat{\alpha}$ in the example reflects the very great difference in the odds of voting for Reagan: Among Democrats, the odds are virtually nil, whereas among Republicans, they are quite large.

This interpretation implicitly assumes fixed marginal totals. But $\hat{\alpha}$ can actually be calculated when none, one, or both sets of marginals are fixed. This is the one measure for which a nil association is 1.0, rather than the customary 0.

Properties of the odds ratio. The odds ratio ranges from 0 to ∞ with 1.0 indicating statistical independence. Values less than 1.0 imply a "negative" association, while values greater than 1.0 mean a positive relationship. In order to see this point, examine these two seemingly different 2×2 tables:

	a			b	
100	50		25	100	
25	200		200	50	
125	250		225	150	

The odds ratios for Tables a and b are 16.0 and .0625, respectively. But notice the similarities: The second table is obtained from the first by simply rotating the frequencies while maintaining the same underlying strength of association. Most observations in Table a lie in the diagonal running from the upper left to the lower right; in the other table they tend to be in the opposite diagonal. In this sense, the two tables reflect similarity in the magnitude but not in the *direction* of the relationship.

Also note that $\hat{\alpha}_b = 1/\hat{\alpha}_a$ (that is, .0625 = 1/16). The upshot is that departures in either direction from 1.0 imply essentially the same thing but are measured on different scales. Negative relationships are measured on the interval 0 to 1.0 and positive relationships on the interval 1.0 to plus infinity. Not being symmetric about 1.0 means that two tables with the same degree of association, but in opposite directions, have different $\hat{\alpha}$'s.

The lack of symmetry[3] is easily removed by calculating the natural logarithm of $\hat{\alpha}$:

$$\hat{\alpha}^* = \log \hat{\alpha} = \log \left(\frac{n_{11} n_{22}}{n_{21} n_{12}} \right)$$

$$= \log n_{11} + \log n_{22} - \log n_{21} - \log n_{12} \qquad [7]$$

The measure $\hat{\alpha}^*$, called the log odds, varies from $-\infty$ to ∞ with 0 indicating independence. In the two previous tables, $\hat{\alpha}^*_a = 2.77$ and

TABLE 13
Effect of Changes in Marginal Distributions on Measures of Association

	a			b	
	First Investigator			Second Investigator	
	X			X	
	75	15		750	15
Y			Y		
	10	100		100	100
Totals	85	115	Totals	850	115

$$\hat{\alpha} = 50 \qquad\qquad \hat{\alpha} = 50$$
$$\hat{\alpha}^* = 3.91 \qquad\qquad \hat{\alpha}^* = 3.91$$
$$\hat{Q} = .96 \qquad\qquad \hat{Q} = .96$$
$$\hat{Y} = .75 \qquad\qquad \hat{Y} = .75$$
$$X^2 = 111.65 \qquad\qquad X^2 = 348.57$$
$$\hat{\Phi}^2 = .56 \qquad\qquad \hat{\Phi}^2 = .36$$
$$r = .75 \qquad\qquad r = .60$$

SOURCE: Reynolds (1977: Table 2.4).
NOTE: Table entries are the number of cases.

$\hat{\alpha}^*_b = -2.77$. Although $\hat{\alpha}^*$ has the appeal of being symmetric, it is perhaps more difficult to interpret than the simple odds ratio.

Both the odds ratio and its logarithm have several important properties. They are invariant under row and column multiplications. To appreciate this feature, consider two hypothetical investigators working on the same problem. Even though their variables are identical, they sample different populations at different rates and thus obtain different marginal distributions (see Table 13).

Not only does the first investigator have a smaller sample (200 versus 965), but the cases are relatively more dispersed on X: The proportions in the first column of each table are .44 and .88, respectively. Yet in spite of these differences, the odds ratio and its logarithm are the same in both tables, namely $\hat{\alpha} = 50$ and $\hat{\alpha}^* = 3.91$. Most social scientists find this stability—a trait not found in many other measures of association— quite useful because the inherent relationships appear essentially equivalent. Indeed, the second table has been obtained from the first simply by multiplying the entries in the first column of Table 13a by 10. As a practical matter, this property allows one to compare relationships across tables drawn from different samples. If the basic form of the association is the same in different populations, the $\hat{\alpha}$'s and $\hat{\alpha}^*$'s will be the same (except for sampling error) no matter how much the marginal distributions vary.

The odds ratio is also invariant under interchanges of rows *and* columns. (Switching only the rows *or* columns changes $\hat{\alpha}$ to $1/\hat{\alpha}$.) In this sense, the odds ratio is a symmetric index.

The odds ratio attains its upper bound under weak perfect association, a fact some statisticians consider a virtue, others a vice. Certainly the form of association exhibited in the two subtables of Table 14 differs substantially.

Letting Y and X denote the row and column variables, respectively, one sees that in the first table, values of Y never occur in the absence of a given value of X, a pattern that does not hold in Table b. There the first column of X is not a good predictor of Y; one can only conclude that a person who is in the second column of X will also be in the second category of Y. In this sense the relationship seems weak.

Yet in the two tables, $\hat{\alpha}$ (and $\hat{\alpha}^*$) equals plus infinity, suggesting an equivalence in relationships. The observed odds ratio equals infinity whenever n_{12} or n_{21} (or both) equal zero. (It equals zero whenever n_{11} or n_{22} or both equal zero.) A similar principle applies whenever one frequency is very small. Changing the 0 in the second table to 1 means $\hat{\alpha}$ = 200, a value many social scientists would still consider misleading.

Yule's Q. One of the best-known measures of association in the social sciences, Yule's Q is a function of the odds ratio and consequently shares most of its strengths and weaknesses. Its definition is

$$\hat{Q} = \frac{n_{11}n_{22} - n_{12}n_{21}}{n_{11}n_{22} + n_{12}n_{21}} = \frac{\hat{\alpha} - 1}{\hat{\alpha} + 1} \qquad [8]$$

For the data in Table 12,

$$\hat{Q} = \frac{(87)(12) - (234)(288)}{(87)(12) + (234)(288)} = -.97$$

Unlike α, Q lies between –1.0 and 1.0 with 0 implying statistical independence. But like the odds ratio, Q attains its upper limit under strict, implicit, *or* weak perfect association. Thus, as was true of $\hat{\alpha}$, $\hat{Q}$ reaches its maximum, 1.0, in both subtables of Table 14. (One readily sees from the appropriate formula that Q is 1.0 whenever n_{12} or n_{21} or both are zero; it is -1.0 whenever n_{11} or n_{22} or both are zero.) For this reason many investigators feel it overstates the strength of an association. Certainly it gives the largest numerical value of all the normed indices usually computed for 2×2 tables, but whether it overstates a relationship depends on one's model of perfect association (Weisberg, 1974). In any event, values close to $|1.0|$ indicate a strong relationship.

Q is invariant under row and column multiplications (see Table 13), and is symmetric.

Yule's Y. Yule's Y, sometimes called the coefficient of "colligation," is also a simple function of the odds ratio:

$$\hat{Y} = \frac{\sqrt{n_{11}n_{22}} - \sqrt{n_{12}n_{21}}}{\sqrt{n_{11}n_{22}} + \sqrt{n_{12}n_{21}}} \qquad [9]$$

For Table 12 the estimate is

$$\hat{Y} = \frac{\sqrt{(87)(12)} - \sqrt{(234)(288)}}{\sqrt{(87)(12)} + \sqrt{(234)(288)}} = -.78$$

Y has the same properties as Q though they are by no means equal in most 2×2 tables. In fact, the absolute value of Y is less than the absolute value of Q except when X and Y are independent or completely associated.

A measure based on chi square, Φ^2. One reason for not using the goodness-of-fit chi square as a measure of association is that its numerical magnitude depends partly on the size of the sample. Dividing chi square by n corrects for n and leads to a popular measure of association—phi squared:

$$\hat{\Phi}^2 = X^2/n$$

where X^2 is the observed goodness-of-fit chi square.[4]

Φ^2 for Table 12 is

$$\hat{\Phi}^2 = \frac{307.719}{621}$$

$$= .50$$

In a 2×2 table, Φ^2 varies between 0 and 1, obviously equaling zero when the variables are statistically independent. It attains its maximum only under strict perfect association. In Table 14, for instance, Φ^2 equals 1.0 in the first subtable but not in the second, where it is only .25.

On the other hand, the marginal variation in X or Y affects its magnitude. As one sees in Table 13 above, the greater the imbalance in the marginal distributions, the lower its value, other things being equal. Using either percentages or the odds ratio as the criterion, the form and strength of association are the same in both tables, but Φ_b^2 is considerably smaller than Φ^2_a. Where one or both marginals are highly skewed, a less sensitive measure may be preferred. Finally, the computation of Φ^2, a symmetric index, does not depend on which variable is considered dependent. The interpretation of Φ^2 can be further facilitated by realizing that it is equivalent to r^2, the square of the product-moment correlation coefficient, applied to a 2×2 table.

The correlation coefficient, r. The categories of dichotomous variables can be coded 0 and 1 and used in the (Pearson) product-moment correlation formula. In a 2×2 table, the calculations reduce to

$$r = \frac{n_{11}n_{22} - n_{12}n_{21}}{\sqrt{n_{1+}n_{2+}n_{+1}n_{+2}}} \qquad [10]$$

Although a symmetric measure, the square of the correlation coefficient is commonly interpreted as the percentage of variation in the dependent variable that is "explained" by the independent variable. The estimate for Table 12 is

$$r = \frac{(87)(12) - (234)(288)}{\sqrt{(321)(300)(375)(246)}} = -.70$$

TABLE 14
Behavior of Measures of Association Under Different Models of Perfect Relationship

		a				b	
		X				X	
		50	0			50	0
Y				Y			
		0	50			50	50
Totals		50	50	Totals		100	50

$\hat{\alpha} = +\infty$ $\hat{\alpha} = +\infty$

$\hat{\alpha}^* = +\infty$ $\hat{\alpha}^* = +\infty$

$\hat{Q} = 1.0$ $\hat{Q} = 1.0$

$\hat{Y} = 1.0$ $\hat{Y} = 1.0$

$\hat{\Phi}^2 = 1.0$ $\hat{\Phi}^2 = .25$

$r = 1.0$ $r = .50$

NOTE: Table entries are the number of cases.

Since $r^2 = .495$, about 49 percent of the variance in voting is accounted for by party identification. Thus, remembering that statistical explanation is not equivalent to theoretical understanding, r has a clear interpretation.

Because r^2 is equivalent to Φ^2 in 2×2 tables, they share the same properties. The correlation coefficient is sensitive to skewed marginal distribution (see Table 13), but is invariant under interchanges of *both* rows and columns. (It changes sign only when the rows or columns are switched.) It is an appropriate measure when one's definition of perfect is strict perfect association (see Table 14).

As is well known, r varies between −1.0 and 1.0. It equals zero if the row and column variables are independent. (It can also equal zero when the variables are *nonlinearly* related.) From the formula it is apparent that $r = 1.0$ if $n_{12} = n_{21} = 0$ and $r = -1.0$ if $n_{11} = n_{22} = 0$. In this sense, the correlation coefficient gives both the direction and strength of association. In a standardized 2×2 table, where each marginal probability equals ½, $r = Y$; otherwise $|r| < |Y|$ except when the variables are independent or completely related. Consequently, its numerical value will usually be less than either Y or Q.

Measures of Association for I × J Tables

Measuring association in I × J tables involves many of the same concepts and problems found in the analysis of 2 × 2 tables. The objective is to find clearly understandable measures, ones that are not confounded by marginal distributions or table dimensions. Although innumerable approaches exist, many of them can be grouped under three headings:

(1) generalizations of the odds ratio,
(2) measures based on chi square, and
(3) "proportional-reduction-in-error" measures.

The odds ratio in I × J tables. The odds ratio or its logarithm readily generalizes to larger tables. An I × J table, where either I or J or both are greater than 2, contains subsets of 2 × 2 tables, and an $\hat{\alpha}$ or $\hat{\alpha}^*$ can be calculated for each. Looking at several individual odds ratios instead of a single summary index permits one to examine various subhypotheses of interest and, in many instances, to locate the precise source of an association.

Let P_{ij} denote the probability that an observation is in the ij[th] cell of an I × J population table. Then a *basic set* of odds ratios is:

$$\hat{\alpha}_{ij} = \frac{P_{ij}P_{IJ}}{P_{iJ}P_{Ij}} , \quad i = 1, 2, \ldots, I-1; j = 1, 2, \ldots, J-1 \qquad [11]$$

Notice that the last (the bottom right hand) cell of the table is the reference point. (Actually, any cell could serve this purpose.) Viewed from this perspective, there are $t = (I - 1)(J - 1)$ 2 × 2 tables in an I × J table, each composed of probabilities from the ith and I[th] rows and the jth and J[th] columns.[5] Corresponding to each such subtable is an odds ratio that can be estimated from a sample by

$$\hat{\alpha}_{ij} = \frac{n_{ij}n_{IJ}}{n_{iJ}n_{Ij}} \qquad [12]$$

where n_{ij} represents the frequency in the ij[th] cell. Of course, the $\hat{\alpha}$'s (or their logarithms), have the same interpretation as odds ratio (or its logarithm) presented earlier.

Using odds ratio to analyze a cross-classification means an investigator must examine a set of coefficients. In a large table the number of possible $\hat{\alpha}$'s will be sizable. Partly for this reason and because it is not a normed index, social scientists have been reluctant to employ this technique in contingency table analysis.

But there are several advantages. As already noted, one can partition a cross-classfication in order to examine various "subhypotheses." In addition to calculating L^2, the likelihood ratio chi square, it is of course possible to compute $\hat{\alpha}$ (or $\hat{\alpha}^*$) for each component table. Table 9 illustrates the point. Predictably, the strongest relationships as measured by $\hat{\alpha}^*$ involve differences between Democrats and Republicans. Notice also that although several L^2's are statistically significant, the strength of the corresponding relationships is relatively small; indeed, only in Tables 6a, 6b, and 7a are the log odds much different than zero. Once again these findings may be obvious, but they demonstrate how odds ratios can be used to pinpoint relationships in a complex table.

Learning to use odds ratios has two other advantages. First, there is a well-developed sampling theory for them so that one can calculate their "simultaneous" confidence intervals. Whenever one makes several significance tests or—what is the same—constructs several confidence intervals on the same set of data, it is necessary to adjust the significance level to take account of the fact that several hypotheses are being tested. Simultaneous inference procedures allow one to make the necessary adjustment. See Reynolds (1977a) or Goodman (1964, 1969) for further details. The second advantage is that the odds ratio lends itself very nicely to the interpretation of log-linear models, a relatively new and important technique for analyzing multidimensional nominal data. (See Knoke and Burke, 1980.)

Measures based on chi square. As in the 2×2 case, chi square alone is not a good indicator of the form or strength of a relationship in a general table, for its magnitude depends partially on n. Standardizing it by dividing by the sample size is an obvious solution. But the resulting measure, Φ^2, does not have an upper bound except in 2×2 tables. Not being bounded, the measure is difficult to interpret. In Table 1, where the goodness-of-fit chi square is 457.634 with 12 degrees of freedom,

$$\hat{\Phi}^2 = 457.634/968 = .47$$

How should this number be interpreted? Since Φ^2 equals zero if the variables are independent, the observed value implies a weak relationship. Without more information, however, one cannot say precisely what .43 means. It is just as hard to do so in tables showing stronger relationships (see Table 15, for example).

Partly for these reasons, a number of normed variations of Φ^2 have been proposed. All of them are symmetric and equal zero when the variables are statistically independent. Two shortcomings are, however, that they frequently cannot attain their maximums, and values lying between 0 and 1.0 are hard to interpret.

The *contingency coefficient*, C, theoretically lies between 0 and 1. For sample data it is estimated by

$$\hat{C} = \sqrt{\frac{\hat{\Phi}^2}{\hat{\Phi}^2 + 1}} = \sqrt{\frac{x^2}{x^2 + n}} \qquad [13]$$

In Table 1,

$$\hat{C} = \sqrt{\frac{457.634}{457.634 + 968}}$$

$$= .57$$

C does not always reach 1.0, even when the variables seem completely associated. In square tables (such as I = J), for instance, its maximum value is $\sqrt{(I-1)/I}$. In this instance one can obtain an "adjusted" C by computing $C_{adj} = C/C_{max}$, where C_{max} is the maximum value C for a particular table.

In asymmetric tables, such an adjustment is less feasible. As an example, Table 15 shows that C is less than 1.0 even though there is an implicit perfect association. Some investigators, furthermore, recommend that C not be used for tables smaller than 5×5 (Garson, 1976).

Another version of Φ^2, Tschuprow's T, varies between 0 (for independence) and 1.0 but can attain its maximum only in square tables. When I does not equal J, T will be less than 1.0 (see Table 15). The sample estimate of T is

$$\hat{T} = \sqrt{\frac{\Phi^2}{\sqrt{(I-1)(J-1)}}} = \sqrt{\frac{x^2}{n\sqrt{(I-1)(J-1)}}} \qquad [14]$$

TABLE 15
Behavior of Measures of Association Under Model
of Implicit Perfect Association

	X				Totals
	0	50	0	50	100
Y	50	0	0	0	50
	0	0	50	0	50
Totals	50	50	50	50	200

$$X^2 = 400$$
$$\hat{\Phi}^2 = 2.000$$
$$\hat{C} = .816$$
$$\hat{T} = .816$$
$$\hat{V} = 1.0$$
$$\lambda_y = 1.0$$
$$\hat{\tau}_y = 1.0$$

NOTE: Table entries are the number of cases.

In Table 1, T is

$$\hat{T} = \sqrt{\frac{457.634}{(968)\sqrt{(2)\,(6)}}} = .37$$

Cramer's V corrects for some of the deficiencies in C and T—it achieves its maximum in asymmetric arrays as in Table 15—but is still rather difficult to interpret. The sample estimate is

$$\hat{V} = \sqrt{\frac{\Phi^2}{m}} = \sqrt{\frac{X^2}{nm}}$$

[15]

where m equals the smaller of $(I - 1)$ or $(J - 1)$. In Table 1, V is

$$\hat{V} = \sqrt{\frac{457.634}{(968)(2)}} = .49$$

Note that V is always at least as large as T.

Besides their sensitivity to table dimensions and marginal distributions, chi-square-based measures do not have intuitively appealing interpretations. Even though they lie between 0 and 1.0 it is hard to understand a value of, say, .49. Presumably the relationship is weak but there is no operational standard for judging its magnitude. These measures were originally intended as crude approximations of the usual correlation coefficient and have been supplemented (if not replaced) by more easily interpreted measures.

Proportional-reduction-in-error measures. To avoid the weaknesses of indices based on chi square, statisticians have developed a variety of other approaches. Perhaps the most popular alternative is proportional-reduction-in-error (PRE) logic.

PRE measures rest on a simple conception of association. Imagine a game in which one randomly draws people from a population and guesses their scores on Y, the dependent variable. The predictions can be made in either of two ways: first, (1) knowing nothing at all about the individuals or (2) knowing their scores on another, independent, variable, X. Whatever rule is followed, one will surely misguess at least some of the time. But if Y depends on X, then knowledge of X categories should reduce the errors.

Each rule has its own probability of error. Under the first rule no information is used to predict scores on Y; the guesses are, in a sense, blind. Denote the probability of misclassifying subjects on this basis by P (A). According to the second rule, one examines each individual's X category and then, based on that information, predicts the value on Y. If each category of X corresponds to one and only one category of Y, then knowing an X category permits guessing the Y category exactly. If, on the other hand, the variables lack perfect correspondence but are related to some degree, the probability of misclassification, P(B), will still be less than under the first rule.

It seems natural to compare the probabilities of making errors under the two rules. To the extent that X is related to Y, the probability of error under the second rule will be less than under the first. The amount of reduction is a criterion for measuring association. Dividing this magnitude by the probability of an error by the first rule gives the *proportional reduction in error*:

$$\frac{\text{PRE Measure}}{\text{of Association}} = \frac{(\text{Probability of Error by Rule 1}) - (\text{Probability of Error by Rule 2})}{(\text{Probability of Error by Rule 1})}$$

$$= \frac{P(A) - P(B)}{P(A)}$$

Suppose, for example, the probability of misclassifying people under Rule 1 on vote in 1980 is .6 but that once information about their party identification is taken into account the probability of making an error drops to .3. The proportional reduction in error is

$$\text{PRE} = \frac{(.6) - (.3)}{(.6)} = .5$$

meaning that knowledge of the independent variable leads to a 50 percent reduction in the expected number of errors that would have been made had party identification not been used.

Properties of PRE measures. Notice first of all that $P(A)$ is always greater than or equal to $P(B)$.[6] If the variables are statistically independent, the categories of X do not supply any information about the Y categories, and $P(A) = P(B)$. In this situation, the PRE measure is zero, as it should be. Unfortunately, there are instances where the measure will be zero though the variables are *not* statistically independent. This problem usually arises when the distribution on Y is highly skewed.

For perfect association, the probability of error under the second rule, $P(B)$, is zero and the measure reduces to

$$\text{PRE Measure} = \frac{P(A) - 0}{P(A)} = 1.0$$

The limits are thus 0 and 1.0—zero when X and Y are independent, 1.0 when they are completely related. The principal advantage is that intermediate values have a clear interpretation as the proportional-reduction-in-error in predicting classes of the dependent variable.

TABLE 16
Example of a Table of Probabilities (hypothetical data)

		X		
		c	d	Totals
Y	a	.3	.1	.4
	b	.2	.4	.6
	Totals	.5	.5	1.0

NOTE: Entries are probabilities. a, b, c, and d are category labels.

At first glance the definition appears somewhat awkward. But the operations are really quite compatible with social science research. After all, most investigators want to predict scores on dependent variables.

Finally, it is clear that one variable must be treated as dependent on the other. Predicting categories of X instead of Y will normally lead to different error rates and hence different values for the PRE measure. In this sense PRE measures are asymmetric: Their numerical value depends on the dependent variable. For the instances when the designation of the dependent variable is arbitrary, one can compute both versions (that is, PRE_Y and PRE_X, where the subscript indicates the dependent variable) or a symmetric version. Because PRE logic is quite broad, numerous nominal measures of association are based on it. Their meaning and computation stem from the precise definition of errors.

Goodman and Kruskal's lambda. Goodman and Kruskal's (1954) lambda rests on very straightforward definitions of prediction error. Referring to the hypothetical population probabilities or proportions presented in Table 16, one sees that the first entry, .3, is the probability of having both characteristic *a* on Y and characteristic *c* on X. The marginal probability, .4 indicates that a member of this population has a four-in-ten chance of being in category *a, regardless of the value on X.*

According to the first rule, one predicts a randomly selected individual's Y class without knowledge of his or her classification on X. How should the prediction be made? The marginal probability of category *b*, .6, is larger than the marginal probability of category *a*, .4.

With no other information available, it would be sensible to guess *b*. Guessing that class *each* time, of course, leads to errors, since not everyone belongs in it. But the probability of *b* is .6, and over the long run 60 percent of the choices should be correct. The proportion of successful predictions by this method exceeds the proportion of successes obtained by predicting *a*. Hence, the first rule is as follows: Always guess the modal class of Y (here *b*) with the probability of error being simply one minus the probability of a success, or

$$1. - .6 = .4.$$

For an I $\times$ J table, let P_{m+} denote the *maximum* marginal row probability. Without knowledge of X, one should always guess the category corresponding to the probability P_{m+}. The probability of making accurate predictions is P_{m+}, while the probability of error is

$$P(A) = 1 - P_{m+}$$

According to the second rule, the investigator selects an individual at random, examines that individual's classification on X, and then predicts the Y category. Again, exactly how should the prediction be made? Continuing with the example, suppose an individual happens to belong to the first category of X in Table 16. The largest cell probability in that column, .3, implies that if a person belongs to category *c* on X he or she is slightly more likely to belong to category *a* than *b* on variable Y. The difference in probabilities suggests predicting category *a* for each individual who has characteristic *c*. Once it is known that a person is in the first column, the best guess is category *a* instead of *b*. Knowledge of the independent variable has altered the prediction made under the first rule. Once again mistakes will be made, because some members of column *c* do not belong in row *a*. Nevertheless, within a given column the errors can be minimized by guessing the most probable row category.

Consider a member of the second column. In what row does the member most likely belong? The largest cell probability in this column, .4, lies in the second row. If the individual's X characteristic is *d*, one would suspect that he or she has characteristic *b*. Predicting *b* for each member of the second column d results in a number of errors, but fewer than if a is chosen.

Finally, to calculate the number of errors under the second rule, add the probability of making errors in each column and subtract from 1.0. Since the probabilities of successful prediction in the two columns are .3 and .4, the probability of error is simply

$$1 - (.3 + .4) = .3$$

(Or conversely, the probability of errors is simply .2 + .1 = .3.)

In an I × J table, the symbol P_{mj} denotes the *maximum* cell probability in the jth column, (In Table 16, P_{m1} and P_{m2} are .3 and .4, respectively.) The probability of error under the second rule, $P(B)$, is

$$P(B) = 1 - \sum_{j=1}^{J} P_{mj}$$

These steps define a PRE measure, lambda:

$$\lambda_y = \frac{(1 - P_{m+}) - \left(1 - \sum_{j=1}^{J} P_{mj}\right)}{(1 - P_{m+})} = \frac{\left(\sum_{j=1}^{J} P_{mj}\right) - P_{m+}}{(1 - P_{m+})} \qquad [16]$$

The subscript Y indicates that categories of Y are being predicted from information about X.

Hence, λ_y and λ_x for the population in Table 16 are

$$\lambda_y = \frac{(.3 + .4) - .6}{(1 - .6)} = .25$$

$$\lambda_x = \frac{(.3 + .4) - .5}{(1 - .5)} = .40$$

Using X as the predictor leads to a 25 percent reduction in error in predicting categories of Y. This reduction suggests a moderately strong relationship between X and Y, at least as the term "relationship" has been defined here. Knowing an individual's X classification does improve predictions of the Y categories. Similarly, in predicting X categories on the basis of Y, the percentage reduction in error decreases by 40 percent.

The data also illustrate the asymmetry of the measures: λ_y does not equal λ_x. One should therefore rely on substantive knowledge to determine the most appropriate index. If, for instance, the aim of a study is the explanation of variation in Y by reference to other variables, λ_y should be computed.

Sample estimates can be calculated by replacing the population probabilities with estimated probabilities. It is simpler, though, to use raw frequencies and compute the estimate of λ_y from the following formula:

$$\hat{\lambda}_y = \frac{\left(\sum_{j=1}^{J} n_{mj}\right) - n_{m+}}{(n - n_{m+})} \qquad [17]$$

where n_{m+} represents the largest marginal row total, n_{mj} represents the largest frequency in the jth column, and n is the sample size. These quantities correspond to the probabilities in the previous expression, and the underlying logic remains the same. The estimate, $\hat{\lambda}_y$, gives the proportional reduction in error for a sample of n observations.

Returning to Table 1, it is easy to estimate the reduction in error in predicting vote based on knowledge of party identification. Since the first row contains the largest marginal total, $n_{m+} = 492$. The sum of the largest cell frequencies in the successive columns is

$$\sum_{j=1}^{7} n_{mj} = 168 + 120 + 49 + 54 + 85 + 131 + 103 = 710$$

Substituting into the previous formula, the estimate for Table 1 is

$$\hat{\lambda}_y = \frac{710 - 492}{968 - 492} = .46$$

Hence, by taking party identification into account, about a 46 percent reduction in error is achieved in predicting candidate preference.

Using an analogous formula, the estimate of λ_x is

$$\hat{\lambda}_x = \frac{(131+168+28) - 201}{(968-201)} = \frac{327 - 201}{968 - 201} = .16$$

From this point of view, the reduction in error is about 16 percent.

Symmetric version of lambda. Whether an investigator wants $\hat{\lambda}_y$ or $\hat{\lambda}_x$ depends on his or her understanding of the variables: if Y depends on X, then $\hat{\lambda}_y$ is appropriate; otherwise $\hat{\lambda}_x$. Occasionally, however, investigators do not know or are unwilling to assume any dependency among the variables. In this case they might prefer a "symmetric" coefficient. Fortunately, slightly modifying the PRE logic, a symmetric version of lambda can easily be defined.

The definition of this measure, denoted simply λ, requires the imaginary process of using the two rules to predict people's classifications. This guessing game, as noted before, is a purely heuristic device designed to clarify the meaning of the measure. One does not actually make any predictions, but only pretends to in order to define association.

Symmetric lambda combines the logic of computing both λ_y and λ_x. Suppose, for example, individuals are randomly selected, and half are assigned to Y classes and half to X categories. According to the first rule, these predictions are made without any additional knowledge. In guessing Y categories, one would always place individuals in the most probable Y class, the one pertaining to P_{m+}, if one wished to minimize the number of errors. On the other hand, when predicting X categories, he would pick the one associated with P_{+m}. During the time Y classes are guessed, the probability of a successful prediction is $\frac{1}{2} P_{m+}$, and during the time X categories are assigned, it is $\frac{1}{2} P_{+m}$. A little thought shows that the probability of an incorrect guess is

$$P(A) = 1 - \frac{1}{2} P_{m+} + \frac{1}{2} P_{+m} = 1 - \frac{P_{m+} + P_{+m}}{2}$$

where p_{m+} and p_{+m} are, respectively, maximum row and column marginal probabilities. (In Table 16, these numbers are .6 and .5.) The factor $\frac{1}{2}$ enters because each probability applies to half of the guesses.

Knowledge of both variables is taken into account under the second rule. For those individuals whose Y classes are being guessed, the investigator uses information about their scores on X. As with λ_y, the probability of a successful guess is a function of the p_{mj}'s. In particular, it is

$$\frac{1}{2} \sum_{j=1}^{J} P_{mj}$$

This is the same probability as before, except that since Y classes are being predicted only half of the time, it is multiplied by one half. For the other individuals—the ones whose X class is being guessed—the probability of a correct prediction knowing Y is

$$\frac{1}{2} \sum_{i=1}^{I} P_{im}$$

The logic is the same as in the calculation of λ_y and λ_x, except that one effectively computes one measure half of the time and the other measure the rest of the time. The probability of error under Rule 2 is then

$$P(B) = 1 - \frac{1}{2} \sum_{j=1}^{J} P_{mj} + \frac{1}{2} \sum_{i=1}^{I} P_{im} = 1 - \frac{\displaystyle\sum_{j=1}^{J} P_{mj} + \sum_{i=1}^{I} P_{im}}{2}$$

PRE logic measures the proportional reduction in error using Rule 2 instead of Rule 1. Hence, symmetric lambda is defined as after algebraic manipulation:

$$\lambda = \frac{\displaystyle\sum_{j=1}^{J} P_{mj} + \sum_{i=1}^{I} P_{im} - P_{m+} - P_{+m}}{2 - P_{m+} - P_{+m}} \qquad [18]$$

For Table 16, symmetric lambda is

$$\lambda = \frac{(.3 + .4) + (.3 + .4) - .6 - .5}{2 - .6 - .5} = \frac{.30}{.90} = .33$$

The sample estimate, computed from observed frequencies, is

$$\hat{\lambda} = \frac{\sum\limits_{j=1}^{J} n_{mj} + \sum\limits_{i=1}^{I} n_{im} - n_{m+} - n_{+m}}{2n - n_{m+} - n_{+m}} \qquad [19]$$

The estimated lambda between party identification and vote in 1980 (Table 1) is thus

$$\hat{\lambda} = \frac{[(168 + 120 + 49 + 54 + 85 + 131 + 103) + (131 + 168 + 28)] - 492 - 201}{2(968) - 492 - 201}$$

$$= .28$$

Goodman and Kruskal's tau. Another PRE measure, Goodman and Kruskal's tau, represents a modification of the hypothetical guessing game. As before, randomly selected individuals are assigned Y scores with and without knowledge of the independent variable. But this time the assignments preserve the original distributions.

Preserving a distribution means that the distribution of guesses is the same as the original distribution. If, for example, n_{1+} and n_{2+} individuals are in the first two categories of Y, then the assignment process will keep exactly n_{1+} and n_{2+} people in those categories. When calculating lambda, everyone is assigned to Y's modal category, thus the pattern of guesses is not the same as the observed distribution. For some purposes it is useful to have a measure based on maintaining the original distribution.

Instead of developing an explicit formula for tau, it is easier to illustrate its computation on the data in Table 1. Suppose 492 individuals are randomly selected from the table and labeled "Reagan voters." Since many of them are not Reagan backers at all, classifying these people in this fashion creates a number of expected errors. How many? Out of 968 respondents, $381 + 95 = 476$ (or $476/968 = .492$) do not belong in that category. Thus the proportion of incorrectly assigned individuals would be .492 or $(.492)(492) = 241.93$ cases.

Now suppose a sample of 381 is drawn from the table and assigned to the "Carter" category. Once again, misclassification would result since

492 + 95 = 587 are not Carter voters, but how many? The probability of error is 587/968 = .606; hence, of the 381 assignments, we would expect about (.606) (381) = 231.04 cases to be incorrectly assigned.

Continuing in this manner, the total number of expected errors under Rule 1—preserving the marginal distribution but not using X to make the predictions—would be

$$= (.492) (492) + (.606) (381) + (.902) (95)$$

$$= 241.93 + 231.04 + 85.68$$

$$= 558.65$$

Notice that in making these guesses the marginal totals have been preserved because 492 cases were classified "Reagan," 381 "Carter," and 95 "Anderson." More generally, to compute the expected errors under Rule 1, subject to the constraint that the marginal totals be maintained, proceed as follows: for each category of the dependent variable, count the number of observations not belonging in it, divide by n, and multiply by the category total. Then sum these totals.

To compute the errors under Rule 2, one continues in exactly the same manner except that it is necessary to work within categories of the independent variable. This amounts to guessing a person's vote preference while knowing his or her party affiliation, but again subject to the constraint of preserving the original distribution, this time within categories.

Consider, for example, the 196 strong Democrats. We have to assign 21 of them to the Reagan category, 168 to Carter, and 7 to Anderson. Suppose we pick 21 of these strong Democrats at random. A few might be Reagan voters, but most will not. Hence, in classifying all of them as Reagan voters, we will surely be making some errors. To find the expected number of errors, first count the number of non-Reagan voters among the strong Democrats: 168 + 7 = 175. Since 175 is about 89 percent of 196, we would expect that of the 21 cases assigned to the Reagan category, about 89 percent or 18.75 cases would be erroneously classified. But next classify 168 of the 196 as Carter voters. Although we will still make a few errors, we will do much better because only 21 + 7 = 28 of the 196 strong Democrats did not vote for Carter. That turns out to be about 14 percent (28/196 = .143) so only about 14 percent of the 168 assignments or (168) (.143) = 24.02 cases will be wrong. And finally, of the 7 people assigned to Anderson,

$$(21 + 168/196) \times 7 = 6.74$$

will be incorrectly classified. Thus, the errors made in assigning the 196 strong Democrats to the three classes of voting total

$$18.75 + 24.02 + 6.74 = 49.51$$

The expected number of errors for the remaining 6 categories of partisanship are found in the same way. The *total* number of errors expected by following the second rule is thus

$$49.51 + 106.57 + 70.39 + 44.75 + 42.98 + 39.05 + 16.91 = 370.16$$

More generally, expected errors from following the second rule are found by taking each level of the independent variable in turn. Working within categories, say the jth, one counts the number of cases not belonging to a specific category of the dependent variable, say the ith, divides this total by n_{+j} and then multiplies by n_{ij}.

The two sets of expected errors are substituted into the PRE formula to obtain an estimate of the proportional reduction in error subject to the constraint that the marginal totals be kept. For Table 1, the estimate is

$$\hat{\tau}_y = \frac{558.65 - 370.16}{558.65} = .34$$

Hence, knowledge of party identification leads to about a 34 percent reduction in error.

A computing formula for sample data is:

$$\hat{\tau}_y = \frac{\sum_i n_{i+} \left[\dfrac{\sum\limits_{\substack{i' \\ i' \neq i}} n_{i'+}}{n} \right] - \sum_j \left[\sum_i n_{ij} \left(\dfrac{\sum\limits_{\substack{i \\ i'=i}} n_{i'j}}{n_{+j}} \right) \right]}{\sum_i n_{i+} \left[\dfrac{\sum\limits_{\substack{i' \\ i' \neq i}} n_{i'+}}{n} \right]} \qquad [20]$$

Here the notation
$$\sum_{\substack{i' \\ i' \neq i}}$$
means that the summation
is taken over each row except the ith. A little reflection will show that this formula simply makes explicit the preceding logic.

Like λ, Goodman and Kruskal's tau lies between 0 and 1: It equals zero if the variables are statistically independent and equals 1.0 under complete association. "Complete" in this concept means "strict" or "implied" perfect association. That is, $\tau = 1$ if for each category of the independent variable, j, there is a category of the dependent variable, i, not necessarily unique, such that $p_{ij} = p_{+j}$. It is also an asymmetric index since $\tau_y \neq \tau_x$ in most cases, as the reader can verify by treating party identification as the dependent variable and calculating tau.

Comparing Measures of Association

A measure of association purportedly describes the type and strength of a relationship between two variables. Ideally, the index tells us how the variables are related in the population, but at the least it should have an unambiguous meaning when applied to observed data. Unfortunately, however, various factors that are not directly related to the nature of the association can affect a measure's numerical value, thereby clouding its interpretation. Perhaps the most important of these factors is the way the variables have been categorized.

Categorization partly affects marginal distributions. The actual distribution of adults in terms of party identification only partially determines their distribution in a sample cross-classification. Question wording and the definition of category boundaries are also important. Suppose, for example, an investigator simply classified people as Democrat, Independent, or Republican, omitting the gradations "strong" and "weak." The resulting variable and marginal distribution would be quite different than what appears in Table 1. Decisions about category boundaries and the like are to some extent arbitrary, because they represent choices rather than the inexorable determination of nature. Consequently, two individuals studying the same phenomenon can obtain different observed marginal distributions. Even if they follow exactly the same operational procedures, they might still get different distributions if they sample populations with different variances. Since measures of association are affected by these distributions, it is entirely possible that their substantive conclusions about the underlying relationship would differ, even though in fact they are the same.

It is important, then, to think carefully about categorization process and marginal distributions.

Indices of dispersion for nominal data.[7] A measure of dispersion or variation indicates the degree of differences among observations on a variable. As opposed to a measure of central tendency, like the mode, it tells how much variation exists among the cases. A useful and simple measure of marginal variation for nominal data is the *index of diversity*:

$$\hat{D} = 1 - \sum_i (n_{i+}/n)^2 \qquad [21]$$

where n_{i+} is the ith marginal total in the ith class and $n = \sum_i n_{i+}$ is the sample size.

The index of diversity gives the probability that a pair of randomly selected observations will be in different categories. If, for example, $\hat{D} = .8$, then the probability that two randomly chosen individuals will have different party identifications is .8. If there is no variation—that is, if all the cases fall in a single category—then $\hat{D} = 0$. If, on the other hand, the variable has I categories, the maximum possible value of $\hat{D}$ is $(I - 1)/I$, which occurs if equal proportions appear in each category. Hence, $\hat{D}$ lies between 0 and $(I - 1)/I$.

Potential variation, it ought to be pointed out, increases as the number of categories increases. Classifying a population into seven groups leads to greater potential variation than simply dichotomizing it: $6/7 = .87$ is of course greater than $1/2 = .5$.

The index of diversity for party identification is

$$\hat{D}_{Party} = 1 - [(196/968)^2 + (201/968)^2 + \ldots + (112/968)^2]$$

$$= .84$$

A related measure is the *index of qualitative variation*:

$$IQV = [I/(I - 1)] \hat{D} = [I/(I - 1)] [1 - \sum(n_{i+}/n)^2] \qquad [22]$$

where I is the total number of categories. The index of qualitative variation, which is a "standardized" version of D because it takes into account the number of categories, has the same interpretation (it is the probability that a randomly selected pair of observations will be in different categories) except that its maximum possible value is 1.0. Thus, $0.0 \leq IQV \leq 1.0$. IQV is appropriate and useful when comparing the variation of distributions based on different numbers of categories. As an example, suppose that a questionnaire administered in the 1950s classified respondents as liberal, conservative, or moderate, whereas a more recent survey used a seven-point scale. It would be preferable to calculate IQV rather than $\hat{D}$ when comparing variation because the number of categories differs.

For Table 1, the index of qualitative variation is

$$IQV = (7/6) (.844) = .98$$

Hence, there is about as much variation (in this sense) as possible, given the nature of categorization.

These measures obviously do not have the same interpretation and mathematical utility as the standard deviation (or variance) calculated on quantitative data. Nevertheless, they can be quite useful in the analysis of nominal scales, as seen below and in later sections.

There has been considerable discussion recently about the decline of political parties. One might suppose among other things that the proportion of the population identifying with a party has decreased while the number of independents has increased. Showing the distribution of partisanship over the last thirty years, Table 17 permits one to examine that supposition. Both $\hat{D}$ and IQV clearly suggest that variation in party identification has changed hardly at all: The probability of two people having the same affiliation is about the same in 1980 as in 1952.

It is also apparent from the table that the way variables are categorized affects the magnitudes of variation. Suppose we decided to collapse party identification into just three categories. The index $\hat{D}$ shows the effect. (Since IQV is a standardized measure—that is, it adjusts for the number of categories—it does not change very much.) For reasons to be seen in a moment, decreasing variation can confuse the interpretation of measures of association, and it is advisable to keep as many categories as possible.

TABLE 17
Distribution of Party Identification, 1952-1980

a

Seven Categories

	1952		1960		1972		1980	
	%	n_{i+}	%	n_{i+}	%	n_{i+}	%	n_{i+}
SD	22	392	20	382	15	404	17	228
D	26	446	26	478	26	700	24	326
ID	10	178	6	115	11	296	11	150
I	6	107	10	191	13	350	12	171
IR	7	125	7	134	11	296	12	171
I	14	250	14	268	13	350	14	193
SR	14	250	16	306	10	269	10	137
Totals:	99	1748	99	1874	99	2665	100	1376
$\hat{D}$:	.82		.83		.84		.84	
IQV:	.96		.96		.98		.98	
$\hat{\sigma}$:	2.20		2.22		1.97		2.00	

b

Three Categories

	1952		1960		1972		1980	
	%	n_{i+}	%	n_{i+}	%	n_{i+}	%	n_{i+}
Democrats	48	838	46	860	41	1104	40	554
Independents	23	410	23	440	35	942	36	492
Republicans	29	500	31	574	23	619	24	330
Totals:	100	1748	100	1874	99	2665	100	1376
$\hat{D}$:	.63		.64		.65		.65	
IQV:	.95		.96		.97		.98	
$\hat{\sigma}$:	.85		.86		.78		.78	

SOURCE: Miller, Miller, and Schneider (1980), Table 2.1, p. 81; 1980 American National Election Study (see Table 1 for complete citation).

KEY: (See Table 3). In Table b, "Democrats" consist of SD and D; "Independents" consist of ID, I, and IR; and "Republicans" of R and SR. Percentages do not add to 100% because of rounding errors.

Finally, note a major difference between these indices and the usual standard deviation, $\hat{\sigma}$. The standard deviation describes the dispersion about a central point, the mean, and one can see that by this definition there has been a very slight decrease in variation. The explanation is, of course, that people are "migrating" from the extreme categories into the middle ones, but the movements are rather even, so that no single category dominates the others. In any event, the change in partisanship is perhaps not so dramatic as commonly supposed, although even marginal changes can greatly affect a party's fortunes at the polls.

How variation and categorization affect measures of association. Generally speaking, the higher the variation, the larger a measure of association *can* be, other things being equal. And as might be expected, if variation is limited by, for example, combining adjacent classes or eliminating categories altogether, an index might be smaller than it would otherwise be. Hypothetical data illustrate these points.

According to most definitions, the variation in a categorical variable is greatest when all categories have equal numbers of cases, as in Table 18a. Combining the last two rows in each table reduces the variation in Y without substantially changing the relationship as measured by percentages. Notice, for example, that 75 percent of the cases lie in the second category of Y in Table 18c, but that the joint percentages are about the same as in Table 18a.

For every decline in Y's variation, each measure except $\hat{\tau}$ also declines. Lambda as well as the indices based on chi square are highest in the symmetric table where variation in Y is greatest and lowest where variation is less. Only tau remains stable through the three tables.

In some tables, in fact, lambda equals zero even if the variables are not statistically independent. This problem arises when the modal class of Y is so large relative to the others that all n_{mj} lie in the same row (see Table 19). As $\hat{\tau}_y$ indicates, X and Y are related, but since the maximum cell frequencies in each column are in the same row,

$$\sum_{j=1}^{J} n_{mj} = n_{m+} \quad \text{and} \quad \hat{\lambda}_y = 0$$

In other tables having a preponderance of cases in one category of Y, lambda may be quite close to zero, suggesting little or no relationship. Tau is less sensitive in this respect and might be used when the dependent variable is highly uneven.

TABLE 18
Effects of Decreasing Variation in the Dependent Variable

a

Cases Distributed Evenly
Among Four Categories of Y
$(\hat{D}_y = .75)$

	X				Totals
	80%	10%	5%	5%	
	(800)	(100)	(50)	(50)	1000
	10	80	5	5	
	(100)	(800)	(50)	(50)	1000
Y	5	5	80	10	
	(50)	(50)	(800)	(100)	1000
	5	5	10	80	
	(50)	(50)	(100)	(800)	1000
Totals	100%	100%	100%	100%	
	(1000)	(1000)	(1000)	(1000)	4000

$x^2 = 6480 \quad \hat{C} = .79 \quad \dot{T} = .90 \quad \hat{\tau}_y = .54$

$\hat{\Phi}^2 = 1.62 \quad \dot{V} = .90 \quad \hat{\lambda}_y = .73$

b

Cases are Unevenly Distributed
$(\hat{D}_y = .62)$

	X				Totals
	80%	10%	5%	5%	
	(800)	(100)	(50)	(50)	1000
Y	10	80	5	5	
	(100)	(800)	(50)	(50)	1000
	10	10	90	90	
	(100)	(100)	(900)	(900)	2000
Totals	100%	100%	100%	100%	
	(1000)	(1000)	(1000)	(1000)	4000

$x^2 = 4520 \quad \hat{C} = .531 \quad \dot{T} = .68 \quad \hat{\tau}_y = .58$

$\hat{\Phi}^2 = 1.13 \quad \dot{V} = .75 \quad \hat{\lambda}_y = .70$

(continued)

Table 18 (Continued)

c

Marginal Distribution in Y is Skewed
$(\hat{D}_y = .37)$

	X				Totals
Y	80% (800)	10% (100)	5% (50)	5% (50)	1000
	20 (200)	90 (900)	95 (950)	95 (950)	3000
	100% (1000)	100% (1000)	100% (1000)	100% (1000)	4000

$x^2 = 2160$ $\hat{C} = .35$ $\hat{T} = .56$ $\hat{\tau}_y = .54$

$\hat{\Phi}^2 = .54$ $\hat{V} = .73$ $\hat{\lambda}_y = .60$

NOTE: Numbers in parentheses are the number of cases. $\hat{D}_y$ is the index of diversity for Y.

TABLE 19

λ_y Equals Zero Even Though There Is a Significant
Relationship Between X and Y

	X			Totals
Y	45% (45)	10% (10)	5% (5)	60
	50 (50)	70 (70)	60 (60)	180
	5 (5)	20 (20)	35 (35)	60
Totals	100% (100)	100% (100)	100% (100)	300

$\hat{\lambda}_y = 0.0$

$\hat{\lambda}_x = .25$

$\hat{\tau}_y = .095$

NOTE: Figures in parentheses are the number of cases.

The same principle applies to changes in the variation of the independent variable. In general, as the variation in X decreases (that is, as its marginal distribution becomes increasingly uneven), most measures decrease, other things being equal. Looking at Table 20, one sees that the basic relationship, as measured by percentages, stays the same in all three subtables. Indeed, the tables are generated by simply halving the frequencies in the first and last columns while keeping the middle column constant. Although hypothetical, the data could represent three samples drawn from populations having different amounts of variation in X.

Whatever the case, the underlying relationship remains the same. But both PRE and chi square measures decline, indicating that the numerical values depend partly on how the cases are distributed. Note, on the other hand, that the basic sets of odds ratios remain constant. (That is, for example, $\hat{\alpha}_{21} = 2.25$ in all three tables.)

Problems of this sort are particularly acute when one is trying to compare tables with unequal marginals. As in the previous section, suppose two investigators who are studying the relationship between X and Y base their analyses on samples from different populations. Calculating either lambda or tau, the first investigator—whose data appear in Table 21a—finds a weak relationship as measured by lambda and tau. The other investigator finds a moderately strong relationship. At first, the results seem incompatible. Yet closer inspection reveals that the relative proportions are exactly the same. (In fact, the entries in the first column of Table 21b have been multiplied by 10 to produce the other data.)

Standardizing a table can often solve problems like this. Standardization is a method for adjusting frequencies to conform to any desired set of marginal totals. An observer may wonder, for example, what the relationship would be if each category of X had the same number of cases. The easiest method is to compute percentages, treating the percentages as though they were raw frequencies. In percentaging on the independent variable, one pretends that each category of X contains exactly 100 cases.

Table 21c is a standardized table. (Since the proportions in both a and b are the same, they generate the same standardized frequencies.) Now the nature of the relationship seems clearer. The first investigator's findings appear somewhat misleading because of the skewed marginals, while results in the second table agree with the standardized data.

TABLE 20
Effects of Decreasing Variation in the Independent Variable

a

Cases Evenly Distributed
Among X Categories
$(\hat{D}_x = .67)$

		X		Totals
	600	300	100	1000
Y	300	400	300	1000
	100	300	600	1000
Totals	1000	1000	1000	3000

$X^2 = 780$ $\hat{V} = .36$ $\hat{\tau}_y = .13$ $\hat{\alpha}_{12} = 6.0$

$\hat{\Phi}^2 = .26$ $\hat{T} = .36$ $\hat{\alpha}_{11} = 36$ $\hat{\alpha}_{22} = 2.67$

$\hat{C} = .45$ $\hat{\lambda}_y = .30$ $\hat{\alpha}_{21} = 6.0$

b

Uneven Distribution
Among X Categories
$(\hat{D}_x = .75)$

		X		Totals
	300	300	50	650
Y	150	400	150	700
	50	300	300	650
Totals	500	1000	500	2000

$X^2 = 406.59$ $\hat{V} = .32$ $\hat{\tau}_y = .10$ $\hat{\alpha}_{12} = 6.0$

$\hat{\Phi}^2 = .20$ $\hat{T} = .32$ $\hat{\alpha}_{11} = 36$ $\hat{\alpha}_{22} = 2.67$

$\hat{C} = .41$ $\hat{\lambda}_y = .23$ $\hat{\alpha}_{21} = 6.0$

(continued)

Table 20 (Continued)

c

Extremely Uneven Distribution
Among X Categories
$(\hat{D}_x = .50)$

		X		Totals
	150	300	25	475
Y	75	400	75	550
	25	300	150	475
Totals	250	1000	250	1500

$X^2 = 211.72$	$\hat{V} = .27$	$\hat{\tau}_y = .07$	$\hat{\alpha}_{12} = 6.0$
$\hat{\Phi}^2 = .14$	$\hat{T} = .27$	$\hat{\alpha}_{11} = 36$	$\hat{\alpha}_{22} = 2.67$
$\hat{C} = .35$	$\hat{\lambda}_y = .16$	$\hat{\alpha}_{21} = 6.0$	

NOTE: Table entries are frequencies. The relative proportions remain the same in every table. D_x is the index of diversity for X.

In general, when comparing tables or when variables are skewed, it will be useful to recompute measures of association on standardized data. This should eliminate to a degree the vagaries of marginal totals. Of course, variation itself may have substantive interest and should also be reported.

It is possible to standardize data so that they conform to any desired set of marginals. Suppose, in particular, that one wanted to know what the relationship between X and Y would be if there were an equal number of cases in the categories of *both* X and Y. A procedure called iterative proportional fitting allows one to adjust the observed frequencies in such a way as to produce the desired marginals *without* changing the nature of the relationship as measured by $\hat{\alpha}$ (or $\hat{\alpha}^*$). Since the calculations are a bit cumbersome, we will not describe them here. (The interested reader should consult Reynolds [1977a: 31-33].) Nevertheless, if one is interested only in the marginal distribution of X, the independent variable, it is sufficient to compute percentages as in Table 21c and treat the percentages as frequencies.

Finally, these remarks suggest two generalizations. First, it usually pays to look at a relationship from several points of view, as each

TABLE 21
Standardizing a Table Helps Remove Effects
of Unequal Marginal Totals

a

First Investigator's Data
$(\hat{D}_x = .28)$

		X		Totals
	350	10	5	365
Y	150	30	15	195
	50	10	35	95
Totals	550	50	55	655

$$\hat{\lambda}_y = .17 \qquad \hat{\tau}_y = .11$$

b

Second Investigator's Data
$(\hat{D}_x = .67)$

		X		Totals
	35	10	5	50
Y	15	30	15	60
	5	10	35	50
Totals	55	50	55	160

$$\hat{\lambda}_y = .40 \qquad \hat{\tau}_y = .21$$

c

Standardized Table
$(\hat{D}_x = .67)$

		X		Totals
	64	20	9	93
Y	27	60	27	114
	9	20	64	93
Totals	100	100	100	300

$$\hat{\lambda}_y = .40 \qquad \hat{\tau}_y = .21$$

NOTE: $\hat{D}_x$ is index of diversity for X.

measure rests on a slightly different definition of association. Unless a theory explicitly assumes a particular definition, which is at most never the case, one may overlook important aspects of the data by relying on a single index.

Second, measures of association by themselves do not prove the relative explanatory power of variables. Social scientists commonly ask for the most important explanation of a given dependent variable. After computing a series of coefficients, it is tempting to take the variable whose index has the largest numerical value as the best predictor or explanation. But since the coefficients are susceptible to extraneous factors like marginal distributions and since they each represent a certain conception of association, these comparisons could be very misleading. In addition, the impact of one variable on another depends partly on its relationship to still other variables, many of which may be unmeasured. For these reasons, using a coefficient of association alone to show explanatory importance seems questionable.

4. INTRODUCTION TO MULTIVARIATE DATA ANALYSIS

Having found a connection between party identification and voting, one wonders if introducing additional variables will further the under-standing of the relationship. The simultaneous examination of more than two variables, called multivariate analysis, raises two general questions. Although it is not possible to explore these problems in detail, a brief introduction may be a useful guide to the more advanced methods.

One may ask in the first place if additional variables improve predictions about the dependent variable. Suppose, for example, that the respondents in Table 1 have been classified by their level of education as well as their partisanship and vote. Does the extra information lead to a better understanding of political preferences?

A second general question concerns changes in a relationship caused by controlling for a third variable. How, for instance, is the association between party identification and vote affected by holding education constant? Table 22 contains three contingency tables, each displaying the association between partisanship and vote *within* a particular level of education. The first table shows the cross-classification among respondents with less than a high school education, the second the relationship among high school graduates, and the third among people with at least some college training. Looking at the data from this perspective, one can ask the following questions:

—Is the nature and strength of the relationship the same in each subtable? If the form of the association varies from one level of the control variable to another, then *interaction* exists. "Interaction" means that the relationship between X and Y is not the same in each category of Z, a control variable (or set of control variables). X and Y might be strongly related in the first level of Z, weakly related in the second, and so forth.

—Assuming no interaction (that is, the relationships are essentially the same in each contingency table), are the variables statistically independent? If so, the *partial association* between X and Y is zero. If, on the other hand, the variables are statistically related and the strength of the relationship is the same across categories of Z, the partial association is constant but not nil.

—If interaction is present, in which subtable is the relationship strongest? In which is it weakest? Is the relationship nil at some levels of Z and not others? The answers *specify* the conditions under which a relationship holds.

Detecting interaction, measuring partial associations, and specifying relationships lie at the heart of many multivariate techniques. As mentioned above, a description of these methods lies beyond the scope of this paper, but the underlying objectives and logic can be outlined.

The Causal Analysis of Nominal Data

In the past, many social scientists recommended "explicating" an original relationship between two variables by holding control or "test" variables constant, thereby creating a series of contingency tables (Hyman, 1955; Rosenberg, 1968). In Table 22, the test factor—education—is the control variable. The objective is to examine the original relationship within levels of the test factor.

What does one look for? The answer depends on what are believed to be the underlying causal mechanisms. A social scientist can never prove causality or even the direction of causation. But if various simplifying assumptions are made, it is at least possible to eliminate models that are inconsistent with the data.

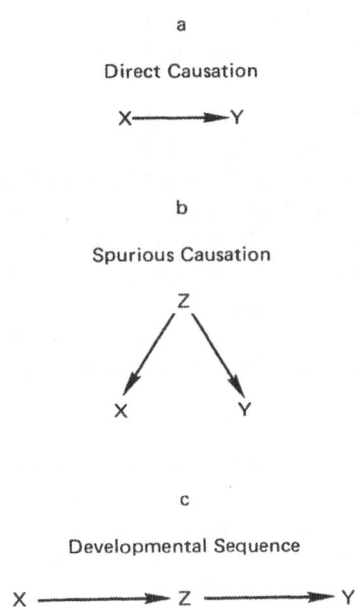

a

Direct Causation

X ————▶ Y

b

Spurious Causation

c

Developmental Sequence

X ————▶ Z ————▶ Y

Figure 1: Three Causal Models

Consider the three models in Figure 1. In the first, X directly causes Y. Changing X will presumably change Y. By contrast, X and Y are "spuriously" related in the second model. A change in X has no effect on Y since both variables are causally dependent on Z. Finally, in the third model, a developmental sequence, X is indirectly linked to Y through Z. Changing X causes Z to change, which in turn produces a change in X.

The second and third models can be distinguished from the first by controlling for Z, the test factor. If the contingent associations between Y and X are nil or at least very much weaker than the original relationships (except for sampling error), then one infers that the relationship is either spurious or a developmental sequence; otherwise the variables are directly related. (Whether it is a spurious or developmental sequence depends on whether Z causally or temporally precedes X—a matter that must be decided on substantive, not statistical grounds.) Even if Y and X are directly related, the nature or form of the association might be affected by Z. With such interaction present, one has to specify the relationship by describing it for each level of Z.

To illustrate these ideas, consider again Table 22. Suppose someone wished to develop a theory explaining variations in political preferences. Past research suggests that party identification is an important determinant but the investigator wants to be sure that the relationship is not spurious, due to education. If it is spurious, the partial associations between candidate preference and partisanship should be zero, except for sampling error. Or, suspecting interaction, one may believe that the relationship will be stronger among high school graduates than college graduates. Assuming that other variables are not confusing the matter, one ought therefore to control for education. Hence, with the proper simplifying assumptions, test factor stratification permits a choice among alternative causal models. There are several ways of carrying out the analysis.

Perhaps the easiest is to "eyeball" each subtable, relying on percentages as a guide. To check for interaction in Table 22, for example, one compares the percentages in all three subtables. In both cases, Democrats are most likely to vote for Carter, and Republicans are more apt to support Reagan. Since the pattern of the relationship seems to be essentially the same, at first sight, interaction does not appear to be present.

It is also clear from an inspection of the tables that controlling for education does not eliminate the basic relationship. The partial correlation, in other words, is not zero. One can determine this by noting the statistically significant chi squares in each table. Thus, neither a spurious relationship nor a developmental sequence seems to hold for these data. (Of course, had another test factor been used, the results might have been different.)

Although these techniques for detecting spuriousness, interaction, and other models have been used for years, they have several shortcomings. One must scan a series of contingency tables. If the control factor has a large number of categories, the number of contingent relationships is quite large. More important, in looking at percentages, particularly in tables with numerous cells, an investigator often has trouble deciding systematically and objectively to what extent the

TABLE 22
Vote by Party Identification by Education

a
Less Than 9 Years

Vote in 1980	Party Identification							Total
	SD	D	ID	I	IR	R	SR	
RR	6% (2)	33% (8)	33% (2)	50% (4)	100% (4)	100% (6)	75% (6)	32
C	94 (32)	67 (16)	50 (3)	50 (4)	0.0 (0)	0.0 (0)	2 (1)	56
A	0.0 (0)	0.0 (0)	17 (1)	0.0 (0)	0.0 (0)	0.0 (0)	13 (1)	2
	100% (34)	100% (24)	100% (6)	100% (8)	100% (4)	100% (6)	100% (8)	90

$X^2 = 50.61$; 12 df

b
9 to 12 Years

Vote in 1980	Party Identification							Total
	SD	D	ID	I	IR	R	SR	
RR	12% (12)	36% (33)	33% (13)	71% (34)	69% (34)	84% (54)	89% (41)	221
C	85 (84)	59 (54)	55 (22)	18 (9)	16 (8)	5 (3)	9 (4)	184
A	3 (3)	4 (4)	12 (5)	10 (5)	14 (7)	11 (7)	2 (1)	32
	100% (99)	99% (91)	100% (40)	99% (48)	100% (49)	100% (64)	100% (46)	437

$X^2 = 180.60$

(continued)

Table 22 (Continued)

c

More Than 12 Years

Vote in 1980	SD	D	*Party Identification* ID	I	IR	R	SR	Total
RR	11% (7)	29% (25)	27% (17)	55% (16)	80% (47)	85% (70)	97% (56)	238
C	82 (50)	58 (49)	38 (24)	21 (6)	9 (5)	5 (4)	0.0 (0)	138
A	7 (4)	13 (17)	35 (22)	24 (7)	12 (7)	10 (8)	3 (2)	67
	100% (61)	100% (91)	100% (63)	100% (29)	101% (59)	100% (82)	100% (58)	443

$$X^2 = 225.68$$

SOURCE: See Table 1.
KEY: (See Table 3).

relationships are nil or to what extent they differ among themselves. The connections between vote and party in the two subtables seem to be the same, but are they? Since the chi squares are so different, one wonders if the percentages are not misleading.

A quick but somewhat more objective way to answer these questions is to compute a coefficient of partial association, proceeding as follows:

(1) stratify the sample by the control variable to create a series of contingency tables (as in Table 22);

(2) compute an index of association for each contingent relationship; and

(3) average the separate indices, the average being interpreted as a measure of partial association.

Either a simple or weighted average can be used. (A weighted average is computed by multiplying each coefficient by the number of cases used in its calculations, adding these multiplications, and dividing the sum by the total number of cases.) Since the two types of averages frequently vary, especially if the numbers of observations in the subtables differ substantially from one another, it is preferable to compute a weighted average whenever the cases are not evenly distributed on the control variable.

In Table 22, for example, where the bulk of the cases are in the subtables b and c, one can compute a weighted average of the lambdas or taus. These averages are interpreted as measures of partial association: They show the "typical" correlation between vote and partisanship after education has been controlled. If the original, uncontrolled relationship is really spurious, then the partial relationships, as measured by lambda or tau, should be approximately zero. Hence, if the average is close to zero, this is evidence of a spurious relationship.

The data, however, do not conform to these expectations, because the estimated partial associations are $\bar{\hat{\lambda}}_y$ = .42 and $\bar{\hat{\tau}}_y$ = .42, where the bar symbol denotes a weighted average. These values do not differ much from the corresponding uncontrolled measures. Within each subtable, there remains an association between vote and party identification, although it is stronger in Tables b and c, suggesting the presence of interaction.

A partial coefficient of this sort has to be interpreted carefully, for these measures—like all indices—are sensitive to how the variables have been categorized. Only if the observed cutpoints closely reflect the variables' true categories can the partials be expected to produce valid and consistent conclusions. Underlying variables are often really continuous or interval-scale, but are dichotomized or trichotomized because the investigator lacks sufficient knowledge to measure them more precisely.

In addition to these rather ad hoc procedures, one can compute more formal partial correlation coefficients for nominal data. Goodman and Kruskal (1954), for example, propose a partial lambda

$$\hat{\lambda}_{yx.z} = \frac{\sum_{k=1}^{K} \sum_{j=1}^{J} n_{mjk} - \sum_{k=1}^{K} n_{m+k}}{n - \sum_{k=1}^{K} n_{m+k}} \qquad [23]$$

where n_{m+k} is the largest row marginal total in the kth subtable and n is the total number of cases in all K subtables. For Table 22 partial lambda is

$$\hat{\lambda}_{yx.z} = \frac{706 - 518}{970 - 518} = .42$$

Other measures of partial association for categorical data are available, but I shall not examine them here. (The interested reader should consult Reynolds [1977].) The difficulty is that even these more sophisticated indices have to be interpreted cautiously.

Unfortunately, crude measurement makes it difficult if not impossible to detect the real underlying model. If Z creates a spurious relationship between X and Y, then the partial association ought to be close to zero or at least substantially smaller than the original, uncontrolled relationship. Yet measures of partial association usually do not behave predictably unless the categorizations represent the true state of affairs.

As a consequence, the investigator cannot be sure that the correct model has been detected. It is imperative, therefore, to think carefully about the level of measurement. Crude measurement usually yields misleading, even erroneous conclusions no matter how sophisticated the technique.

5. CONCLUSIONS

The procedures described in this paper have been greatly augmented by recent developments in the analysis of categorical data. Two are worth mentioning.

Log-linear models allow one to express the logarithm of cell frequencies (or some other function of these frequencies) as a function of a linear model. The parameters (or absence of parameters) of such models often have interesting substantive interpretations. Consider as an example the data in Table 22. One might postulate the following model:

$$\log (F_{ijk}) = \mu + \mu_i^V + \mu_j^P + \mu_k^E \qquad [24]$$

where F_{ijk} is the *expected* frequency in the ijk^{th} cell and log refers to the natural logarithm. The specific definition of these terms is not important, however. What is worth noting is that models of this sort have a particular theoretical interpretation. Equation 24, for instance, asserts that the three variables vote, partisanship, and education are all mutually independent; that is, each pair of variables is statistically unrelated.

Presumably this model would not fit the data. To be sure, one could obtain estimated expected frequencies under the model ($\hat{F}_{ijk}$) and compare them with the corresponding observed counts (n_{ijk}), using the usual goodness-of-fit chi square. (If X^2 is large, the model would be rejected.) An alternative might be

$$\log (F_{ijk}) = \mu + \mu_i^V + \mu_j^P + \mu_k^E + \mu_{ij}^{VP} + \mu_{ik}^{VE} + \mu_{jk}^{PE} \qquad [25]$$

In other words, this model asserts that although all three variables are related, there is no "interaction"; that is, the nature of the relationship between, say, voting and party identification, is the same in all levels of education. (See the discussion in the preceding section.)

Under the right circumstances, log-linear models lead to a more thorough understanding of the data than can be gained from simple measures of association. Unfortunately, we cannot go into the details here. Suffice it to say that a good place to begin to understand them is the material presented earlier on odds ratios and on the multivariate analysis of nominal data. For further references see Reynolds (1977) and Knoke and Burke (1980).

The second general approach worth mentioning, however briefly, applies mainly to two-way tables but can be applied to multi-way tables as well. It is sometimes possible and desirable to estimate "association" models for a table. Consider once more the data in Table 1, which consists of 3 rows and 7 columns. One can compute a set of $(I-1)(J-1) = (3-1)(7-1) = 12$ basic odds ratios as follows:

$$\alpha_{ij} = \frac{F_{ij} \, F'_{i+1,j+1}}{F_{i+1,j} \, F_{i,j+1}} \qquad \begin{array}{l} i = 1, 2 \ldots I-1 \\ j = 1, 2 \ldots J-1 \end{array} \qquad [26]$$

(See Goodman, 1979.)

These odds ratios define the relationship between vote and partisanship for a specific 2×2 subtable formed by considering rows i and i+1 and columns j and j+1. The set of α's pertaining to the 12 subtables thoroughly describes the relationship between the two variables. (Actually, we could consider other basic sets of tables; see Goodman [1969].)

Indeed, it is possible to express α's themselves as functions of various parameters. One possibility is

$$\alpha_{ij} = \theta \qquad\qquad [27]$$

This model asserts that the relationship between the two variables is the same or uniform throughout the table. An alternative model might be that the nature of the relationship depends on the nature of the particular rows and columns; that is, $\alpha_{ij} = \Theta_{i+}\Theta_{+j}$.

Once again, it is not possible to go into this approach in detail. Yet, since it promises to provide rich insights into categorical data, particularly ordinal variables, the reader is encouraged to pursue the matter in greater detail. (See Goodman, 1979, 1981.)

NOTES

1. The data, the 1980 American National Election Study, were made available by the Inter-university Consortium for Political Research of the Institute for Social Research, University of Michigan. The consortium is not responsible for any errors or interpretation of these data.

2. Actually, the data in Table 1 and in the other tables in this paper do not come from a simple random sample. They were generated by a more complex sampling procedure. But since they only illustrate various statistics, they are treated as if they were a random sample.

3. Note that the term "symmetry" does not have the same meaning as in Chapter 2. The earlier discussion of symmetry referred to whether or not a measure of association depended on defining a dependent variable.

4. The formula for chi square in a 2×2 table is

$$X^2 = \frac{n(n_{11}n_{22} - n_{12}n_{21})^2}{n_{1+}n_{2+}n_{+1}n_{+2}}$$

where n_{i+} and n_{+j} are marginal totals ($i = 1,2; j = 1,2$).

5. There are actually more than t subtables, but the remainder can be generated from this basic set (see Goodman, 1969).

6. A PRE measure is undefined whenever $p(A) = 0$. This contingency will not arise, however, because if there is no possibility of misclassification, the subjects must all occupy the same Y category and data in this form—a $I \times J$ table—would not be analyzed by these methods.

7. This discussion is based on Agresti and Agresti (1977).

REFERENCES

AGRESTI, A. and B. F. AGRESTI (1977) "Statistical analysis of qualitative variation," in K. F. Schuessler (ed.) Sociological Methodology 1978. San Francisco: Jossey-Bass.

BLALOCK, H. M., Jr. (1979) Social Statistics. New York: McGraw-Hill.

COCHRAN, W. G. (1954) "Some methods of strengthening the common χ^2 tests." Biometrics 10: 417-451.

GARSON, G. C. (1976) Political Science Methods. Boston: Holbrook.

GOODMAN, L. A. (1981) "Three elementary views of log linear models for the analysis of cross-classification having ordered categories," in S. Leinhardt (ed.) Sociological Methodology 1981. San Francisco: Jossey-Bass.

———(1979) "Simple models for the analysis of association in cross-classification having ordered categories." Journal of the American Statistical Association 74 (September): 537-552.

———(1969) "How to ransack social mobility tables and other kinds of cross-classification tables." American Journal of Sociology 75 (July): 1-40.

———(1964) "Simultaneous confidence limits for cross-product ratios in contingency tables." Journal of the Royal Statistical Society (Series B) 26: 86-102.

———and W. H. KRUSKAL (1954) "Measures of association for cross-classifications." Journal of the American Statistical Association 49 (December): 732-764.

HABERMAN, S. J. (1973) "The analysis of residuals in cross-classified tables." Biometrics 29: 205-220.

HYMAN, H. H. (1955) Survey Design and Analysis. Glencoe, IL: Free Press.

IVERSON, G. R. (1979) "Decomposing chi square." Sociological Methods and Research 8: 143-157.

KNOKE, D. and P. BURKE (1980) Log-Linear Models. Beverly Hills, CA: Sage.

MILLER, W. E., A. H. MILLER, and E. J. SCHNEIDER (1980) American National Election Studies Data Sourcebook, 1952-1978. Cambridge, MA: Harvard University Press.

REYNOLDS, H. T. (1977a) The Analysis of Cross-Classifications. New York: Free Press.

———(1977b) "Some comments on the causal analysis of surveys with log-linear models." American Journal of Sociology 83 (March): 127-143.

ROSENBERG, M. (1968) The Logic of Survey Analysis. New York: Basic Books.

WEISBERG, H. (1974) "Models of statistical relationships." American Political Science Review 68 (December): 1638-1655.

H. T. REYNOLDS, Professor of Political Science at the University of Delaware, graduated from Dartmouth College and holds a Ph.D. from the University of North Carolina. His research interests are survey research methods and political behavior. He is currently working on a contextual study of American elections.

In compliance with GPSR, should you have any concerns about the safety of this product, please advise: International Associates Auditing & Certification Limited The Black Church, St Mary's Place, Dublin 7, D07 P4AX Ireland EUAR@ie.ia-net.com